CURRENT ISSUES IN INDIAN AGRICULTURE

CURRENT ISSUES IN INDIAN AGRICULTURE

Edited by
Jagannath Lenka

SERIALS PUBLICATIONS
NEW DELHI (INDIA)

First Published – 2010

ISBN: 978-81-8387-255-3

Published by

SERIALS PUBLICATIONS
4800/24, 1st Floor, Ansari Road
Darya Ganj, New Delhi-110002 (India)
Phone : 23245225. Fax : 91-11-23272135
E-mail: serials@satyam.net.in

Dedicated to

MY FATHER

Late Gouranga Chandra Lenka

CONTENTS

PREFACE

"India is essentially Rural India and Rural India is virtually the cultivator, the village handicraftsman and the agricultural labourer" (AIRCS).Rural India where 70 per cent of all Indians live still depends heavily on agriculture. Its forward and backward linkages with other sectors of the economy are well established in the development economics literature. Needless to say that since independence our overarching focus has been and will continue to be on the role of agricultural development in overcoming backwardness. But unfortunately the sector has performed much worse during the last decade.

The present volume is an attempt to bring to the fore some of the inherent and recent issues in the agriculture sector. Particularly it covers agricultural issues relating to farmers' suicide, employment, growth, productivity, technology, diversification, finance and market. Included in this volume are papers most of which were presented in the 44th Annual conference of the Indian Econometric society.

I am extremely thankful to all esteemed contributors for having made their articles available for this volume.

I wish to record my heartfelt gratitude to my colleagues Dr.Nakulananda Mohanty, Reader in Zoology and Dr (Mrs.) Minati Mallick for their constant encouragement and help extended to me for this work.

I am extremely thankful to my wife Smt. Manoja Manjari Parida and my son Rudra Pravat for their inspiration.

Last but not the least; thanks are due to Serials Publications, Delhi for taking active interest in the publication of this book.

J. Lenka

LIST OF CONTRIBUTORS

Bimal Kishore Sahoo, Department of HSS IIT, Roorkee.

Braja Bandhu Swain, Research Scholar Centre of Development Studies, Tiruananthapuram, Kerala.

C. H. Shah, Sri Sathya Sai University Vidyagiri, Prasanthi Nilayam Anantapur, Andhra Pradesh.

Falguni Pattnaik, Research Scholar Department of HSS, IIT, Kharagpur, West Bengal.

G. Balachandran, Sri Sathya Sai University Vidyagiri, Prasanthi Nilayam Anantapur, AndhraPradesh.

Harish Mani Sri Sathya, Sai University Vidyagiri,Prasanthi Nilayam Anantapur, Andhra Pradesh.

Jagannath Lenka, Professor of Economics North Orissa University Baripada, Orissa, India.

K. P. Mani, Professor of Economics, Dr John Mathai Centre University of Calicut Aranattukara,Trichur Kerala.

Kabita Kumari Sahu, Lecturer in Economics, North Orissa University Baripada, Orissa, India.

Kuldeep, Research Scholar Department of Economics University of Rajasthan, Jaipur.

Lopamudra Mishra, P.G. Department of Economics, Samablpur University Jyoti Vihar, Sambalpur, Orissa.

Minati Mallick, Lecturer in Economics North Orissa University Baripada, Orissa, India.

Narayan Chandra Nayak, Associate Professor of Economics Department of HSS, IIT, Kharagpur, West Bengal

Nilabja Ghosh, Institute of Economic Growth University enclave Delhi.

Raju Mandal, Lecturer in Economics, Assam University Silchar, Assam.

S. Natchathira, Reader in Economics P.G & Research Department of Economics Lady Doak College, Madhurai-2

S. S. Somra, Associate Professor of Economics Department of Economics University of Rajasthan, Jaipur.

Sangeet Chakarvarty, Institute of Economic Growth University Enclave, Delhi.

Shailesh Kumar, Institute of Economic Growth University enclave Delhi.

Srijit Mishra, Associate Professor of Economics IGIDR, General A.K.Vaidya Marg, Goregaon (E), Mumbai, India.

Sudhakar Patra, Reader in Economics F.M.University Balasore,Orissa.

Surath Nayak, Faculty, Department of Economics Bharathidarsan Governmnet College (Autonomous) M.G. Road,V.T.of Puducherry.

Tushar Kanti das, Department of Business Administration Sambalpur University Jyoti Vihar, Sambalpur, Orissa.

V. Pandit, Professor of Economics Sri Sathya Sai University Vidyagiri, Prasanthi Nilayam Anantapur, AndhraPradesh.

1

AGRARIAN SCENARIO IN POST REFORM INDIA: A STORY OF DISTRESS, DESPAIR AND DEATH

Srijit Mishra

INTRODUCTION

In recent years, the growth of the Indian economy at about eight per cent per annum is considered as a success of the reform process initiated in the 1990s. This is analogously called as the rise of the Indian elephant which, of course, is different from the Chinese dragon. Is this poetic phrase justified; or, is it a satire? The grace and beauty in the elephant's stride seems to be missing–"India shines, but Bharat falters." The impressive growth in recent years is largely a story of the urban-based service sector and to a lesser extent for industry whereas agriculture is lagging behind. Today, agriculture accounts for less than one-fourth of the gross domestic produce but employs nearly three-fifths of the total work force. Within agriculture, the incremental value addition in output indicates a shift away from traditional crops to high value crops like fruits & vegetables that hardly have any presence under the gross cropped area. The growth of the cereals, propelled largely by rice and wheat through the green revolution, is also not very encouraging in the recent past.

Overall, income from cultivation is inadequate. It becomes difficult for the farmer to plan for all possible risks: vagaries of nature (primarily, inadequate or excessive water), market related uncertainties such as increasing input costs and output price shocks, unavailability of credit from institutional sources or excessive reliance on informal sources with a greater interest burden and new technology among others. With the decline in extension service he has to rely on the input dealer leading to supplier-induced-demand. This has adverse implications on the livelihoods of the cultivators, most of whom are marginal and small farmers, as well as for agricultural labourers. This is indicative of a larger agrarian crisis. Response to the crisis would be different among different sub-groups and vary across different regions. One of the extreme forms of reaction is reflected through the increasing incidence of farmers' suicides. If the state of farmers is critical then that of the agricultural labourers would be worse off.

In this paper we further elaborate about the changing agrarian scenario, increasing incidence of farmers' suicides and also on some public policy interventions with regard to employment programmes in sections 2-4 respectively. Concluding remarks are in section 5.

CHANGES IN AGRARIAN CONDITIONS[1]

Agriculture's contribution to the gross domestic produce in India has reduced from 56 per cent in 1950-51 to 23 per cent in 2005-06 whereas as per the 2001 census 58 per cent of the total work force and 73 per cent of the rural workers are still dependent on agriculture. This also indicates that rural non-farm employment opportunities are limited. Between 1960-61 and 2002-03, the number of agricultural operational holdings in rural India during Kharif nearly doubled from 53 million to 102 million. Composition across size-class after including landless indicates that large (above 10 hectares), medium (4-10 hectares) and semi-medium (2-4 hectares) categories have been declining from the 1960s, small (1-2 hectares) size-class started declining from the 1970s with absolute numbers declining in the 1990s, marginal (0.002-1 hectares) size-class declined in the 1990s with absolute numbers increasing from 45 million in 1990-91 to 48 million in 2002-03 and nil (0-0.002 hectares) category which declined from 27 per cent in 1970-71 to 22 per cent in 1990-91 has increased to 32 per cent in 2002-03 (National Sample Survey Organisation (NSSO) 2006). After excluding Jammu & Kashmir and Manipur the census figures indicate that from 1991 to 2001 the total number of cultivators remained around 125 million whereas that of agricultural labourers increased from 86 million to 106 million. In short, the dependence on agriculture is increasingly among the ranks of agricultural labourers and marginal farmers.

In the 1990s (1990-91 to 1999-2000) the index of farm income remained around 100 whereas consumer price index of agricultural labourers more than doubled from 100 to 219 and this has led to widening disparities between agricultural and non-agricultural income (Reddy 2006). The index of agricultural production increased from 100 at triennium ending 1981-82 to 170 in the triennium ending 2003-04 (Reserve Bank of India (RBI) 2005). Comparing trends in growth rate in production, area and yield in the 1980s (TE 1981-82 to TE 1992-93) and 1990s (TE 1993-94 to TE 2004-05) for major crop groups one observes that the growth rates have been significantly lower in production and this is largely because of the lower growth rates in yield for most of the crop groups (Table 1).

Using slightly different periodization, Singh (2006) observes similar differences in the annual compound growth rates for the two periods. He also indicates that in the 1980s the growth in value of output per hectare at 2.2 per cent per annum was slightly higher than costs 1.8 per cent per annum whereas in the 1990s the growth in value of output at 0.9 per cent per annum was lower than that of costs at 1.2 per cent per annum and as a result farm business income in recent years has not increased much. Incremental value addition in agriculture (TE 2002-03 over TE 1992-93) shows a shift towards high value addition crops like fruits and vegetables (Appendix 1). A calculation for Maharashtra indicates that increase in value of output in recent years is largely for fruits and vegetables that account for less than 5 per cent of the gross

States/Union Territories	Incremental Value, (1993-94 prices, Rs Lakh)	Share of Incremental Value											
		Cereals	Pulses	Oilseeds	Fibres	Sugar	Fruits and Vege-tables	Condi-ments and Spices	Drugs and Narcotics	By Pro-ducts	Other Crops	Kitchen Garden	Per cent
Mizoram	7307	32.6	-5.2	-2.1	-1.1	1.6	60.6	24.0	-6.9	-0.7	-3.5	0.7	100.0
Nagaland	52155	12.0	2.6	7.3	0.1	-0.9	60.8	16.4	0.2	0.9	0.1	0.4	100.0
Orissa	-67335	70.6	65.9	71.1	1.8	10.3	-131.5	4.6	3.3	6.6	-0.2	-2.5	100.0
Pondicherry	645	-5.9	-19.9	-67.8	-7.3	-36.6	227.8	-9.3	14.0	-3.7	7.0	1.5	100.0
Punjab	169019	121.0	-2.8	-6.8	-41.2	8.6	19.6	1.1	0.1	1.5	-0.9	-0.1	100.0
Rajasthan	-8338	-436.4	291.8	282.8	417.0	70.3	-262.9	-468.1	-45.2	100.7	147.6	2.3	100.0
Sikkim	621	3.0	-56.8	-76.9	0.0	0.0	206.4	41.1	0.0	-15.7	0.0	-1.2	100.0
Tamil Nadu	151842	-28.0	-4.8	-12.7	-9.7	35.6	97.4	13.9	5.2	0.2	1.4	1.6	100.0
Tripura	37769	14.3	-0.1	-1.8	-0.5	-0.6	68.4	7.3	5.5	-0.4	7.5	0.4	100.0
Uttar Pradesh	689337	46.4	-5.1	-3.9	-0.1	12.6	44.8	0.2	2.3	2.3	-0.2	0.7	100.0
West Bengal	536327	28.8	-0.1	2.0	4.6	1.0	51.0	1.6	9.9	0.9	-0.1	0.4	100.0
India	4124466	22.1	-1.6	-1.3	-0.8	9.8	57.8	7.2	4.5	-0.2	2.0	0.5	100.0

Note and Source: Author's calculation based on raw data provided by EPWRF, original source of data is Ministry of Statistics and Programme Implementation, Government of India.

cropped area. Evidence from Telengana, Andhra Pradesh indicate that between 1986-2000 real agricultural output growth is significant whereas at the same time there have been significant welfare declines not only for agricultural labourers and marginal farmers, but also for other groups; growth and distress seem to be intertwined (Vakulabharanam, 2005).

In 2002-03, the average returns from cultivation per hectare in India are Rs. 6756/- in Kharif and Rs. 9290 in Rabi (Appendix 2). From the total farmer households, 86 per cent with an average land size of 1.2 hectares in Kharif and 62 per cent with an average land size of 0.9 hectare in Rabi had cultivated. Paid out expenses as per cent of value of output is about 44 per cent in Kharif and 42 per cent in Rabi. This is likely to be higher if one includes imputed family labour or rent on account of own land. Besides, some pattern could be hidden because the calculations are aggregated across all crops. There is wide inter-state variation. Compared to the national average, one observes relatively lower returns per hectare and greater share of expenses in the states of Andhra Pradesh, Gujarat, Haryana, Karnataka, Maharashtra, Madhya Pradesh, Orissa, Rajasthan and Tamil Nadu during Kharif. This could be indicative of high costs or crop failure. Share of expenses to the value of output is less than 30 per cent in most of the hilly states: Himachal Pradesh, Jammu and Kashmir, Jharkhand, the Northeast states and Uttaranchal indicating that dependence on market based inputs could be much lower here.

Average returns from cultivation is Rs. 11,259/- per annum (Table 2).[2] About 60 per cent and 10 per cent of farmer households obtain some returns from farm animals and non-farm business respectively and the monthly returns from these per farmer households are Rs. 85/- and Rs. 236/- respectively. In addition, farmer households will also get income from wages and salaries. As expected, returns per household increases with land size. Average family size also increases with land size indicating that the increase in per capita returns would not be as large. Across caste groups, scheduled castes (SCs, who generally own the marginal lands) have the least returns and above them are scheduled tribes and from both

Appendix 2a
Returns to Cultivation Across States, 2002-03

States	Kharif				Rabi					
	Farmer HHs Cultiva-, ting, %	*GCA per Cultivating HH, Ha*	*Gross returns per Ha, Rs*	*Expenses by value of output, %*	*Farmer HHs Cultiva-ting, %*	*GCA per Cultiva-, ting HH, Ha*	*Gross returns per Ha, Rs*	*Expenses by value of output, %*	*Returns per Farmer HH, Rs*	*Average Family Size*
Andaman & Nicobar Is	98.0	0.9	17928	14.8	70.6	0.6	15952	5.9	21889	5.0
Andhra Pradesh	81.7	1.2	5243	62.3	39.1	0.9	7815	52.8	8105	4.7
Arunachal Pradesh	74.8	1.3	13909	13.5	68.4	0.8	8433	22.1	17950	5.0
Assam	95.9	0.8	16257	12.7	84.7	0.4	16089	19.0	19145	5.7
Bihar	87.0	0.7	8065	39.6	95.2	0.6	10180	37.8	10635	6.1
Chandigarh	59.6	0.9	15120	40.6	68.8	0.5	20782	34.5	15715	6.3
Chattisgarh	98.6	1.3	5355	39.2	26.9	0.8	4296	39.4	7732	5.4
Dadra&Nagar	99.3	0.6	5424	37.5	30.0	0.7	5052	44.2	4455	5.5
Daman&Diu	89.3	0.2	8007	43.7	39.0	0.2	5375	54.7	1677	5.6
Delhi	82.9	0.5	25848	43.5	75.9	0.5	30814	32.6	23834	5.1
Goa	78.0	0.6	11911	27.1	47.0	0.7	15992	27.9	10304	4.5
Gujarat	85.2	1.6	6005	46.8	39.3	1.2	8621	49.3	12527	5.5
Haryana	64.1	1.6	5832	56.9	64.0	1.6	14537	40.7	21055	6.0
Himachal Pradesh	97.1	0.5	16432	28.2	94.6	0.4	5377	50.8	9536	5.2
Jammu & Kashmir	94.7	0.7	28446	17.7	84.9	0.6	10833	26.8	24759	5.8
Jharkand	96.9	0.7	10420	21.1	41.8	0.3	14117	28.1	8577	5.4
Karnataka	95.1	1.5	6522	46.0	47.1	1.3	6536	36.6	13395	5.3
Kerala	93.6	0.4	17724	38.8	94.5	0.4	18220	34.1	13521	4.7
Lakshadweep	80.8	0.1	82751	7.8	100.0	0.1	42229	9.2	14530	6.8
Madhya Pradesh	77.3	1.6	3882	45.3	67.3	1.7	7305	35.8	13146	5.8

Contd...

States	Kharif				Rabi					
	Farmer HHs Cultiva-, ting, %	GCA per Cultivating HH, Ha	Gross returns per Ha, Rs	Expenses by value of output, %	Farmer HHs Cultiva- ting, %	GCA per Cultiva-, ting HH, Ha	Gross returns per Ha, Rs	Expenses by value of output, %	Returns per Farmer HH, Rs	Average Family Size
Maharashtra	94.4	1.6	6609	45.0	45.2	1.1	5505	47.9	12721	5.1
Manipur	84.3	0.6	16697	28.2	51.3	0.2	6682	41.9	9093	5.5
Meghalaya	99.5	1.0	22860	18.1	96.1	1.3	11082	17.9	36415	5.4
Mizoram	90.9	1.0	18905	3.8	89.3	1.1	14823	3.5	32793	5.7
Nagaland	91.2	0.5	29592	7.2	96.3	0.3	17578	17.1	20105	5.1
Orissa	98.1	0.8	3633	48.1	25.2	0.4	5284	50.9	3307	5.0
Pondicherry	47.4	1.3	11883	50.0	38.9	0.7	9421	57.7	9783	4.7
Punjab	36.9	2.5	19974	38.4	27.6	2.6	20929	37.5	32866	5.9
Rajasthan	91.3	1.9	271	89.0	37.0	1.4	10954	40.5	5972	5.9
Sikkim	98.9	0.7	11807	22.3	97.6	0.4	7275	33.8	10900	5.3
Tamil Nadu	79.1	0.8	6682	57.9	44.1	0.8	8562	45.1	7511	4.4
Tripura	91.3	0.5	15333	30.5	62.7	0.3	17500	31.0	9231	5.0
Uttar Pradesh	81.8	0.7	7025	44.1	87.8	0.9	8490	46.0	10824	6.1
Uttaranchal	93.1	0.5	36647	11.9	93.0	0.4	8914	28.0	19315	5.1
West Bengal	89.7	0.5	10942	44.6	71.0	0.4	10976	57.2	8174	5.3
India	86.2	1.1	6756	43.9	62.3	0.9	9290	42.2	11258	5.5

Note:. GCA=Gross Cropped Area, Ha=Hectares, HH=Household, Rs=Indian Rupees, Gross returns equal value of output minus expenses. Expenses are those paid out only, and hence, it does not include family labour or rent for own land. Value of output is total produce times price, it includes by-products also. Source: Calculated from unit level data using 33rd Schedule, 59th round of National Sample Survey (NSS) on Situation Assessment Survey of Farmers.

Appendix 2b (1)
Returns to Cultivation across NSS Regions, 2002-03

NSS Regions	*Kharif*				*Rabi*					
	Farmer HHs Cultivating,%	*GCA per Cultivating HH, Ha*	*Gross returns per Ha, Rs*	*Expenses by value of output, %*	*Farmer HHs Cultivating, %*	*GCA per Cultivating HH, Ha*	*Gross returns per Ha, Rs*	*Expenses by value of output,%*	*Returns to Cultivation per Farmer HH,Rs*	*Average Family Size*
AN1	98.0	0.9	17928	14.8	70.6	0.6	15952	5.9	21889	5.0
AP1	68.6	0.9	6813	66.1	44.8	0.8	5890	62.8	6662	4.4
AP2	94.7	1.3	5087	57.9	34.2	0.9	9331	43.4	9072	4.8
AP3	84.5	2.0	5448	54.0	35.4	1.3	9018	39.6	13366	5.3
AP4	76.7	0.9	-855	108.6	41.1	0.9	8651	62.5	2604	4.5
AR1	74.8	1.3	13909	13.5	68.4	0.8	8433	22.1	17950	5.0
AS1	97.7	0.9	17573	9.8	86.6	0.4	14900	17.8	19469	5.6
AS2	94.1	0.8	14858	16.0	84.1	0.5	16920	19.7	19149	5.7
AS3	100.0	0.7	20211	7.0	74.4	0.2	11922	19.0	16136	5.4
BI1	88.9	0.6	6699	42.0	96.2	0.6	10022	38.2	8996	5.9
BI2	83.9	0.8	9909	37.1	93.5	0.7	10393	37.3	13347	6.4
CN1	59.6	0.9	15120	40.6	68.8	0.5	20782	34.5	15715	6.3
CT1	98.6	1.3	5355	39.2	26.9	0.8	4296	39.4	7732	5.4
DD1	89.3	0.2	8007	43.7	39.0	0.2	5375	54.7	1677	5.6
DE1	82.9	0.5	25848	43.5	75.9	0.5	30814	32.6	23834	5.1
DN1	99.3	0.6	5424	37.5	30.0	0.7	5052	44.2	4455	5.5
GO1	78.0	0.6	11911	27.1	47.0	0.7	15992	27.9	10304	4.5
GU1	93.8	1.0	6454	29.1	30.6	0.7	13474	34.4	9102	5.9
GU2	75.9	1.1	3639	58.4	52.0	1.1	9088	47.0	8363	5.2
GU3	85.4	1.2	9488	41.5	37.0	1.2	7179	57.6	12651	5.9
GU4	84.6	2.4	2793	62.3	41.3	1.8	10229	48.2	13377	5.7
GU5	89.5	2.9	7507	45.7	30.6	1.4	4504	62.3	21564	5.3
HA1	65.7	1.3	8854	55.6	69.1	1.4	16504	41.8	23095	6.0
HA2	62.2	1.9	3075	59.9	57.6	2.0	12601	39.2	18515	5.9

Contd...

	Kharif				Rabi					
NSS Regions	Farmer HHs Culti-vating,%	GCA per Cultivating HH, Ha	Gross returns per Ha, Rs	Expenses by value of output, %	Farmer HHs Culti-vating, %	GCA per Cultivating HH, Ha	Gross returns per Ha, Rs	Expenses by value of output,%	Returns to Cultivation per Farmer HH,Rs	Average Family Size
HP1	97.1	0.5	16432	28.2	94.6	0.4	5377	50.8	9536	5.2
JH1	96.9	0.7	10420	21.1	41.8	0.3	14117	28.1	8577	5.4
JK1	87.5	0.8	16234	19.0	80.9	0.8	9918	26.8	17523	5.5
JK2	100.0	0.9	8029	20.1	93.1	0.9	11068	18.2	16989	5.2
JK3	94.4	0.5	67803	17.0	80.7	0.3	11360	41.5	34119	6.3
KA1	95.6	0.8	12171	43.7	80.6	0.5	14538	29.3	14874	6.0
KA2	98.7	1.3	8472	47.4	42.0	1.0	6893	39.8	13696	4.8
KA3	95.3	0.9	6827	51.3	40.2	0.9	6322	43.4	8114	4.9
KA4	93.2	2.2	5538	43.3	49.2	1.9	5945	34.8	16956	5.7
KE1	91.3	0.5	14788	39.3	94.0	0.5	15048	31.5	14345	5.4
KE2	95.0	0.3	20438	38.5	94.8	0.3	21278	35.8	13015	4.3
LA1	80.8	0.1	82751	7.8	100.0	0.1	42228	9.2	14530	6.8
MA1	97.1	0.5	7030	45.6	10.4	0.6	8365	37.9	3747	4.9
MA2	88.7	1.1	10070	41.6	54.6	1.0	5448	53.0	12475	5.1
MA3	94.3	1.9	6776	45.7	34.4	1.4	10460	37.6	17318	5.8
MA4	95.9	2.2	7252	37.7	57.7	1.3	4046	43.5	18339	5.5
MA5	98.6	2.3	4494	53.3	48.0	1.1	5773	51.0	13278	4.7
MA6	98.8	1.5	2968	63.6	33.9	0.6	1843	64.1	4812	4.8
ME1	99.5	1.0	22860	18.1	96.1	1.3	11082	17.9	36415	5.4
MI1	90.9	1.0	18905	3.8	89.3	1.1	14823	3.5	32793	5.7
MN1	76.7	0.6	13673	47.9	55.3	0.1	664	91.5	6199	5.7
MN2	90.8	0.6	18736	11.7	47.9	0.2	11621	20.0	11539	5.3
MP1	80.3	1.2	3799	29.8	81.8	1.5	6750	30.3	11828	5.9
MP2	44.4	1.7	1736	68.7	93.7	2.1	5447	43.3	11860	5.5
MP3	97.1	1.9	4190	43.7	35.7	1.2	11788	32.2	12765	6.1
MP4	93.0	1.5	4607	37.2	71.0	1.4	7176	33.3	13281	5.2
MP5	94.5	2.5	5237	53.1	69.5	2.0	9503	36.2	25565	5.6
MP6	57.2	1.4	1865	45.8	51.3	1.9	8619	34.5	9799	6.1

Appendix 2b (2)
Returns to Cultivation Across NSS Regions, 2002-03

	Kharif				Rabi					
NSS Regions	Farmer HHs Cultivating, %	GCA per Cultivating HH, Ha	Gross returns per Ha, Rs	Expenses by value of output, %	Farmer HHs Cultivating, %	GCA per Cultivating HH, Ha	Gross returns per Ha, Rs	Expenses by value of output, %	Returns to Cultivation per Farmer HH, Rs	Average Family Size
NA1	91.2	0.5	29592	7.2	96.3	0.3	17578	17.1	20105	5.1
OR1	98.1	0.7	4387	52.2	44.5	0.4	4917	49.5	3904	5.3
OR2	97.0	0.9	2482	48.6	10.0	0.8	4142	46.6	2590	4.5
OR3	98.7	0.8	3755	42.2	14.3	0.3	8503	56.3	3115	4.9
PO1	47.4	1.3	11883	50.0	38.9	0.7	9421	57.7	9783	4.7
PU1	35.8	2.3	18584	40.4	20.0	2.2	16684	42.2	22600	5.7
PU2	38.2	2.6	21312	36.7	36.5	2.8	23038	35.6	44813	6.0
RA1	88.4	4.2	-304	121.3	27.6	2.8	9817	38.4	6433	6.0
RA2	87.8	1.2	-327	111.3	49.4	1.2	11003	44.2	6153	6.4
RA3	98.6	0.7	3520	30.6	28.8	0.4	12500	27.6	3593	5.0
RA4	98.0	1.2	3182	50.4	32.7	0.7	18383	33.9	7921	5.5
SI1	98.9	0.7	11807	22.3	97.6	0.4	7275	33.8	10900	5.3
TN1	74.4	0.7	7300	61.4	34.3	0.7	8952	48.5	6198	4.8
TN2	74.6	0.6	2402	80.5	40.4	0.9	5822	59.8	3225	4.5
TN3	80.3	0.8	3998	65.1	35.2	0.7	6925	45.7	4242	4.0
TN4	88.1	1.1	10865	44.1	69.1	1.0	10748	35.5	17615	4.2
TR1	91.3	0.5	15333	30.5	62.7	0.3	17500	31.0	9231	5.0
UP1	82.4	0.8	9903	43.1	85.3	0.8	14110	46.3	16784	6.1
UP2	72.0	0.6	8574	36.9	80.3	1.5	4910	45.5	9784	5.6
UP3	87.4	0.7	5055	48.9	92.9	0.6	7202	47.0	6869	6.3

Contd...

contd.

NSS Regions	*Kharif* Farmer HHs Cultivating, %	*Kharif* GCA per Cultivating HH, Ha	*Kharif* Gross returns per Ha, Rs	*Kharif* Expenses by value of output, %	*Rabi* Farmer HHs Cultivating, %	*Rabi* GCA per Cultivating HH, Ha	*Rabi* Gross returns per Ha, Rs	*Rabi* Expenses by value of output, %	Returns to Cultivation per Farmer HH, Rs	Average Family Size
UP4	67.9	1.2	1248	46.8	88.5	1.7	7385	42.1	12135	5.7
UT1	93.1	0.5	36646	11.9	93.0	0.4	8914	28.0	19315	5.1
WB1	92.7	0.6	15583	23.6	95.9	0.4	17775	41.3	15082	5.4
WB2	85.4	0.6	11185	44.5	81.5	0.5	10237	51.4	9934	5.2
WB3	86.1	0.5	9528	54.2	70.6	0.4	7612	72.4	5850	5.2
WB4	97.0	0.5	10316	42.8	52.6	0.3	14255	52.5	6827	5.4
India	86.2	1.1	6756	43.9	62.3	0.9	9290	42.2	11258	5.5

Note and Source: As in Appendix 2a. The region codes indicate the following: AN1=Andaman & Nicobar Is, AP=Andhra Pradesh, AP1=Coastal, AP2=Inland Northern, AP3=South Western, AP4=Inland Southern, AR1=Arunachal Pradesh, AS=Assam, AS1=Plains Eastern, AS2=Plains Western and AS3=Hills, BI=Bihar, BI1=Northern, BI2=Central, CN1=Chandigarh, CT1=Chattisgarh, DD1=Daman & Diu, DE1=Delhi, DN1=Dadra & Nagar Haveli, GO1=Goa, GU=Gujarat, GU1=Eastern, GU2=Plains Northern, GU3=Plains Southern, GU4=Dry Areas, GU5=Saurashtra, HA=Haryana, HA1=Eastern, HA2=Western, HP1=Himachal Pradesh, JK=Jammu & Kashmir, JK1=Mountainous, JK2=Outer Hills, JK3=Jhelam Valley, JH1=Jharkhand, KA=Karnataka, KA1=Coastal & Ghats, KA2=Inland Eastern, KA3=Inland Southern, KA4=Inland Northern, KE=Kerala, KE1=Northern, KE2=Southern, LA1=Lakshadweep, MA=Maharashtra, MA1=Coastal, MA2=Inland Western, MA3=Inland Northern, MA4=Inland Central, MA5=Inland Eastern, MA6=Eastern, ME1=Meghalaya, MI1=Mizoran, MN=Manipur, MN1=Plains, MN2=Hills, MP=Madhya Pradesh, MP1=Vindhya, MP2=Central, MP3=Malwa, MP4=South, MP5=South Western, MP6=Northern, NA1=Nagaland, OR=Orissa, OR1=Coastal, OR2=Southern, OR3=Northern, PO1=Pondicherry, PU=Punjab, PU1=Northern, PU2=Southern, RA=Rajasthan, RA1=Western, RA2=North Eastern, RA3=Southern, RA4=South Eastern, SI1=Sikkim, TN=Tamil Nadu, TN1=Coastal Northern, TN2=Coastal, TN3=Southern, TN4=Inland, TR1=Tripura, UT1=Uttaranchal, UP=Uttar Pradesh, UP1=Western, UP2=Central, UP3=Eastern, UP4=Southern, WB=West Bengal, WB1=Himalayan, WB2=Eastern Plains, WB3=Central Plains and WB4=Western Plains.

these groups the other backward castes fare better, but the returns for all these three groups is lower than the overall average. Overall, there is not much diversification and the income of an average farmer household from cultivation would hardly suffice to meet some basic day-to-day requirements.

Table 2
Returns to Cultivation, Farm Animals and Non-Farm Business for Farmer Households, 2003

Sub-groups	*Prop of farmer HH, %*	*Returns from Kharif per annum, Rs*	*Returns from Rabi per annum, Rs*	*Returns from Farm Animals (30 days), Rs*	*Returns from Non Farm Busi-ness (30 days), Rs*	*Average Family Size*
Land Size						
Near landless	9.9	367	462	125	339	5.0
Marginal	55.6	3243	2667	88	223	5.2
Small	18.1	8098	5922	100	181	5.7
Semi-Medium	10.6	13880	10596	69	188	6.2
Medium	4.8	22841	20940	75	422	6.9
Large	0.9	33494	34600	122	507	7.5
Caste						
SC	17.4	3123	2693	23	213	5.4
ST	13.3	6256	2746	79	138	5.3
OBC	41.5	5237	5044	92	238	5.6
Others	27.6	9559	7695	140	293	5.5
Like Farming as a Profession						
No	40.1	4156	3337	71	263	5.5
Yes	59.5	7606	6237	103	213	5.5
Total	100.0	6200	5059	85	236	5.5

Note: HH=Household, Rs=Indian Rupees. Near landless=0-0.099 hectares (ha), Marginal=0.1-1 ha, Small=1.001-2, Semi-Medium=2.001-4 ha, Medium=4.001-10 ha, Large>10 ha. Information on caste and whether they like farming as a profession was not available for 0.1 per cent and 0.4 per cent respectively. Returns to Kharif and Rabi are calculated by subtracting paid out expenses from the value of output, which includes by-products. It does not include family labour or rent for own land. Returns from farm animals and non-farm business are calculated based on 30 day recall. The farmer households will also have other sources of income like wages and salaries.

Source: Calculated from unit level data using 33rd schedule, 59th round of National Sample Survey (NSS) on Situation Assessment Survey of Farmers.

At a time when one would require greater inputs from the state, it seems to be withdrawing. This is evident from decline in public investment, in the reduced role of formal credit institutions and poor extension service among others. We elaborate on these. Gross fixed capital formation in agriculture as a proportion of gross domestic product (GDP) declined from 3.1 per cent during 1980-85 (Sixth plan) to 1.6 per cent during 1997-2002 (Ninth plan) in 1993-94 prices (Table 3). During the same period, gross fixed capital formation in agriculture as a proportion of total gross fixed capital formation declined from 13.1 per cent to 7.4 per cent and proportion of plan expenditure towards agriculture & allied activities declined from 6.1 per cent to 4.5 per cent.

Table 3
Capital Formation and Plan Expenditure in Agriculture

Year	*GFCF in Agr as % of GDP, India*	*GFCF in Agr. as % of total GFCF, India*	*Exp on Agr & Allied as % of total Plan Exp., India*
1980-85, Sixth Plan (Actuals)	3.1	13.1	6.1
1985-92, Seventh Plan (Actuals)	2.3	9.6	5.9
1992-97, Eighth Plan (Actuals)	1.9	7.4	5.1
1997-2002, Ninth Plan (Actuals)	1.6	7.4	4.5

Note: GFCF indicates Gross Fixed Capital Formation, GDP indicates Gross Domestic Product at Factor Cost, Exp indicates expenditure, Agr indicates Agriculture.

Source: Mishra (2006a).

In recent years, using 1999-2000 prices one observes that gross fixed capital formation in agriculture has declined from 2.2 per cent in 1999-2000 to 1.7 per cent in a quick estimate of 2004-05. The demand and use of water in domestic consumption (particularly, in urban areas) is also on the rise. These have an adverse impact on irrigation and consequently on availability of water for agricultural purposes. In other words, it is the marginal and small farmers who bear the brunt of unavailability of water and its associated yield uncertainty.

During the 1970s and 1980s the banking industry through affirmative measures led to deployment of credit in agricultural sector. As a result, the scheduled commercial banks share of credit to agriculture increased from 9 per cent in December 1972 to nearly 18 per cent by December 1987 and thereafter the share of agricultural credit declined. The increase in 1970s and 1980s was largely realized through the earmarking of credit for priority sector lending and that too through direct advances. Over time, there was some dilution as it allowed for achieving one-fourth of the target in the form of indirect advances. Despite this, both direct and indirect credit declined to 11 per cent by March 2004. Concurrently, the number of agricultural loan accounts in scheduled commercial banks that had reached a peak of 27.7 million by March 1992 declined to 20.3 million by March 2002 and is at 21.3 million by March 2004 (Shetty 2006).

The Situation Assessment Survey of Farmers of 2003 (NSS 59th round) indicates that nearly 49 per cent of farmer households are indebted with the average outstanding amount per indebted farmer household being Rs. 25902/-. The states with a higher average outstanding amount of indebtedness are Andhra Pradesh, Chandigarh, Gujarat, Haryana, Himachal Pradesh, Karnataka, Kerala, Madhya Pradesh, Maharashtra, Pondicherry, Punjab, Rajasthan and Tamil Nadu (Figure 1). These are also the states with a relatively higher proportion of farmer households indebted, and hence, amount outstanding per farmer household is also higher than the all India average in all these states except for Himachal Pradesh (Appendix 3).

Purpose wise distribution indicates that nearly 58 per cent of outstanding debt is for agricultural purpose (31 per cent capital expenditure and 28 per cent current expenditure), 7 per cent for non-farm business and the remaining 35 per cent is for consumption or other purposes (NSSO 2005). The states with relatively higher proportion of outstanding debt for current expenditure are Andhra Pradesh, Chattisgarh, Gujarat, Karnataka, Maharashtra, Meghalaya and Punjab. Some of the Northeast states and other hilly states indicated that more than 10 per cent of the outstanding debt was for non-farm business; of course the amount of outstanding debt in these states was much lower. One exception is Himachal Pradesh where amount of outstanding debt is higher and 29 per cent of the outstanding debt is for non-farm business.

Appendix 3a
Some Aspects of Indebtedness in Farmer Households, 2003

States	Far HH in Lakhs	Out Amt of Far HH in Rs. Crore	Prop of Far HH Indebted, %	Avg Out Amt per Far HH in Rs	Prop of Out Amt for Prod purposes, %	Prop of Out Amt from Formal sources, %	Prop of Out Amt >1 Yr, %	Prop of SM Far HH to all Far HH, %	Prop of SM Far HH Indebted, %	Out Amt of SM Far HH to Out Amt of all Far HH, %	Out Amt of SM Far HH from Formal sources, %
Andaman & Nicobar Is	0.1	5.7	26.6	5079	56.6	74.0	94.2	77.3	21.2	83.5	14.1
Andhra Pradesh	60.3	14460.0	82.1	23965	64.7	30.6	67.6	78.1	81.5	66.4	19.9
Arunachal	1.2	6.1	5.9	493	10.6	26.9	30.2	69.4	6.1	71.8	1.9
Assam	25.0	203.5	18.1	813	39.4	37.4	51.0	90.0	18.4	83.1	6.9
Bihar	70.8	3168.9	33.0	4476	47.1	41.7	76.3	93.4	33.9	87.0	11.9
Chandigarh	0.0	4.0	46.4	19917	39.8	36.7	59.3	95.1	45.7	45.9	1.6
Chattisgarh	27.6	1137.6	40.2	4122	78.4	71.3	63.6	77.3	39.1	48.0	25.9
Dadra&Nagar	0.2	4.0	26.4	1736	49.0	51.4	84.4	93.1	24.5	87.1	15.2
Daman&Diu	0.0	1.5	41.4	3117	68.4	61.4	61.7	99.0	41.5	99.1	23.1
Delhi	0.1	7.6	42.9	6702	33.4	89.3	65.8	98.8	43.4	100.0	46.1
Goa	0.6	7.8	6.6	1292	39.8	64.3	100.0	94.7	3.8	94.0	3.9
Gujarat	37.8	5875.9	51.9	15526	74.5	68.8	65.4	75.9	46.1	39.5	23.0
Haryana	19.4	5056.9	53.1	26007	69.0	67.2	60.8	76.1	49.2	51.9	31.3
Himachal Pradesh	9.1	871.4	33.4	9618	48.5	63.5	61.2	93.1	33.0	82.7	20.7
Jammu & Kashmir	9.4	179.5	31.8	1903	53.2	67.5	50.9	91.6	30.1	77.4	19.5
Jharkand	28.2	622.5	20.9	2205	57.3	64.1	82.4	94.7	21.0	87.5	12.9
Karnataka	40.4	7328.9	61.6	18135	78.0	68.8	67.9	74.2	60.7	51.8	34.8
Kerala	21.9	7441.4	64.4	33907	44.2	81.4	73.8	96.7	64.4	90.3	52.8
Lakshadweep	0.0	1.1	47.3	6183	2.7	34.4	56.5	99.1	46.8	99.7	14.9

Contd...

States	Far HH in Lakhs	Out Amt of Far HH in Rs. Crore	Prop of Far HH Indebted, %	Avg Out Amt per Far HH in Rs	Prop of Out Amt for Prod purposes, %	Prop of Out Amt from Formal sources, %	Prop of Out Amt >1 Yr, %	Prop of SM Far HH to all Far HH, %	Prop of SM Far HH Indebted, %	Out Amt of SM Far HH to Out Amt of all Far HH, %	Out Amt of SM Far HH from Formal sources, %
Madhya Pradesh	63.2	8986.7	50.8	14218	69.7	56.3	77.6	66.7	45.2	40.0	22.4
Maharashtra	65.8	11171.3	54.8	16973	80.2	83.0	74.3	68.6	48.8	44.9	39.9
Manipur	2.1	48.7	24.8	2269	15.7	18.2	87.6	98.6	24.8	97.1	4.8
Meghalaya	2.5	1.8	4.0	72	78.5	6.0	33.2	81.6	4.4	95.6	0.2
Mizoram	0.8	14.6	23.6	1876	66.5	63.6	78.2	83.7	25.1	96.8	15.8
Nagaland	0.8	8.3	36.5	1030	36.4	68.8	11.3	95.1	37.7	85.8	24.5
Orissa	42.3	2485.7	47.8	5871	64.7	74.3	72.3	92.3	47.1	86.5	34.5
Pondicherry	0.3	63.7	80.1	20927	27.6	59.6	52.6	85.6	82.0	85.6	46.0
Punjab	18.4	7667.4	65.4	41576	66.7	47.5	55.8	75.7	59.8	29.1	26.3
Rajasthan	53.1	9752.1	52.4	18372	59.4	33.7	79.8	66.0	50.5	41.6	12.9
Sikkim	0.5	10.9	32.8	2053	39.2	57.8	50.5	95.1	33.3	88.7	19.6
Tamil Nadu	38.9	9316.8	74.5	23963	54.9	51.4	71.2	87.5	74.9	72.9	34.4
Tripura	2.3	69.5	49.2	2977	59.1	79.7	89.8	99.5	49.4	100.0	39.4
Uttar Pradesh	171.6	12738.9	40.3	7425	67.9	59.8	67.0	90.7	39.2	68.9	19.6
Uttaranchal	9.0	99.3	7.2	1108	51.5	76.1	60.1	96.2	7.0	98.5	9.5
West Bengal	69.2	3625.5	50.0	5237	55.9	57.7	64.5	96.9	50.1	92.6	28.0
India	893.5	112445.4	48.6	12585	65.1	57.0	69.7	83.6	46.3	58.9	23.8

Note: Far=Farmer, HH=Household, Out Amt=Outstanding Amount, Avg=Average, Rs=Indian Rupees, Prop=Proportion, Yr=Year, SM=Small and Marginal.

Source: Calculated from unit level data using 33rd Schedule, 59th round of National Sample Survey (NSS) on Situation Assessment Survey of Farmers.

Appendix 3b (1)
Some Aspects of Indebtedness in Farmer Households, 2003 (NSS Regions)

NSS Regions	Far HH in Lakhs	Out Amt of Far HH in Rs. Crore	Prop of Far HH Indeb-ted, %	Avg Out Amt per Far HH in Rs	Prop of Out Amt for Prod purposes, %	Prop of Out Amt from Formal sources, %	Prop of Out Amt >1 Yr, %	Prop of SM Far HH to all Far HH, %	Prop of SM Far HH Indeb-ted, %	Out Amt of SM Far HH to Out Amt of all Far HH,%	Out Amt of SM Far HH from Formal sources,%
AN1	0.1	5.7	26.6	5078.5	56.6	74.0	94.2	77.3	21.2	83.5	14.1
AP1	23.5	5285.9	84.7	22522.6	61.4	30.0	62.2	86.8	83.9	69.6	20.3
AP2	23.8	5173.7	79.7	21735.1	69.1	31.5	68.3	74.8	78.2	67.1	17.9
AP3	7.7	1660.2	81.7	21586.2	64.3	32.2	64.0	58.7	80.3	41.0	21.4
AP4	5.4	2340.3	82.2	43534.1	62.5	28.7	80.7	82.8	84.6	75.9	22.4
AR1	1.2	6.1	5.9	493.4	10.6	26.9	30.2	69.4	6.1	71.8	1.9
AS1	10.6	69.7	15.6	659.5	54.5	30.8	43.0	87.9	16.5	86.2	6.4
AS2	13.3	131.8	20.3	989.6	31.8	41.5	55.4	91.0	20.0	81.2	8.0
AS3	1.2	2.0	15.8	171.9	19.9	0.0	43.0	97.8	16.1	100.0	0.0
BI1	44.1	1630.4	35.0	3694.4	47.6	42.6	78.4	94.3	36.2	92.9	13.4
BI2	26.7	1538.5	29.7	5768.1	46.5	40.7	74.1	92.0	30.0	80.7	9.5
CN1	0.0	4.0	46.4	19917.1	39.8	36.7	59.3	95.1	45.7	45.9	1.6
CT1	27.6	1137.6	40.2	4121.9	78.4	71.3	63.6	77.3	39.1	48.0	25.9
DD1	0.0	1.5	41.4	3116.5	68.4	61.4	61.7	99.0	41.5	99.1	23.1
DE1	0.1	7.6	42.9	6702.3	33.4	89.3	65.8	98.8	43.4	100.0	46.1
DN1	0.2	4.0	26.4	1735.6	49.0	51.4	84.4	93.1	24.5	87.1	15.2
GO1	0.6	7.8	6.6	1292.0	39.8	64.3	100.0	94.7	3.8	94.0	3.9
GU1	8.4	788.3	39.8	9404.2	72.7	68.9	82.6	89.3	38.0	52.1	18.9
GU2	11.2	1383.4	48.6	12348.3	67.8	58.3	73.6	87.5	46.6	60.9	18.7
GU3	5.6	928.2	50.4	16492.0	59.4	54.8	81.2	83.1	48.8	42.6	18.1
GU4	4.8	417.1	43.4	8650.3	70.9	63.3	57.7	64.2	39.4	42.5	22.8
GU5	7.8	2358.9	75.9	30206.4	85.6	81.3	49.9	47.1	63.3	21.0	45.0
HA1	10.8	2645.1	50.6	24534.2	73.4	70.1	62.9	80.9	47.6	54.2	29.3
HA2	8.7	2411.7	56.2	27839.1	64.1	64.1	58.4	70.2	51.5	49.5	34.0

Contd...

NSS Regions	Far HH in Lakhs	Out Amt of Far HH in Rs. Crore	Prop of Far HH Indeb-ted, %	Avg Out Amt per Far HH in Rs	Prop of Out Amt for Prod purposes, %	Prop of Out Amt from Formal sources, %	Prop of Out Amt >1 Yr, %	Prop of SM Far HH to all Far HH, %	Prop of SM Far HH Indeb-ted, %	Out Amt of SM Far HH to Out Amt of all Far HH,%	Out Amt of SM Far HH from Formal sources,%
HP1	9.1	871.4	33.4	9617.6	48.5	63.5	61.2	93.1	33.0	82.7	20.7
JH1	28.2	622.5	20.9	2204.6	57.3	64.1	82.4	94.7	21.0	87.5	12.9
JK1	2.1	121.7	12.4	5801.1	65.7	87.7	57.1	90.1	12.2	72.5	10.4
JK2	3.1	28.5	80.6	914.1	37.1	26.1	25.6	84.7	81.2	85.3	21.7
JK3	4.2	29.3	5.4	694.8	17.2	23.4	50.2	97.4	5.5	90.1	0.9
KA1	2.8	997.0	56.1	35708.7	78.8	91.9	90.7	88.0	54.5	87.0	51.6
KA2	7.8	1758.2	79.5	22442.5	79.1	61.3	64.1	79.3	77.5	33.6	23.2
KA3	12.7	1808.0	63.1	14201.9	65.8	41.9	67.9	87.5	64.6	72.8	23.8
KA4	17.1	2765.6	53.2	16215.5	85.1	82.8	62.2	59.8	47.7	37.1	33.5
KE1	8.3	2513.9	62.7	30130.5	48.1	88.8	71.9	95.1	62.2	88.5	55.8
KE2	13.6	4927.4	65.4	36223.7	42.3	77.7	74.8	97.7	65.7	91.3	51.5
LA1	0.0	1.1	47.3	6182.9	2.7	34.4	56.5	99.1	46.8	99.7	14.9
MA1	7.2	186.6	30.5	2587.6	69.0	69.9	72.9	92.3	29.3	74.9	20.3
MA2	20.2	4615.9	64.1	22840.7	73.4	79.4	80.1	79.1	62.1	60.3	49.0
MA3	7.1	1775.7	52.1	24961.4	80.7	85.6	64.8	68.2	45.9	43.2	41.8
MA4	13.8	2346.7	52.6	16956.0	88.8	86.4	79.4	56.5	42.6	27.5	34.4
MA5	11.5	1663.4	59.9	14482.7	85.2	83.9	59.3	49.7	48.7	29.5	43.4
MA6	6.0	583.1	51.8	9786.2	87.3	92.0	79.5	69.7	44.1	33.0	42.4
ME1	2.5	1.8	4.0	72.1	78.5	6.0	33.2	81.6	4.4	95.6	0.2
MI1	0.8	14.6	23.6	1875.7	66.5	63.6	78.2	83.7	25.1	96.8	15.8
MN1	1.0	43.4	41.1	4415.2	16.5	19.9	89.7	97.6	41.4	96.8	8.7
MN2	1.2	5.3	11.1	454.4	9.8	3.9	70.5	99.6	11.1	100.0	0.6

Appendix 3b (2)

Some Aspects of Indebtedness in Farmer Households, 2003 (NSS Regions)

NSS Regions	Far HH in Lakhs	Out Amt of Far HH in Rs Crore	Prop of Far HH Indebted, %	Avg Out Amt per Far HH in Rs	Prop of Out Amt for Prod purposes, %	Prop of Out Amt from Formal sources, %	Prop of Out Amt >1 Yr, %	Prop of SM Far HH to all Far HH, %	Prop of SM Far HH Indebted, %	Out Amt of SM Far HH to Out Amt of all Far HH, %	Out Amt of SM Far HH from Formal sources, %
MP1	13.8	614.1	31.5	4442.6	61.9	56.1	83.8	70.1	27.8	45.9	13.8
MP2	11.8	1739.1	56.9	14756.5	70.6	65.3	79.9	67.6	46.1	39.5	25.9
MP3	13.5	3698.4	78.9	27385.1	62.5	42.1	78.7	66.1	76.6	40.2	27.9
MP4	10.3	659.3	32.0	6372.8	85.7	86.4	79.2	70.5	28.5	27.4	28.6
MP5	5.2	453.8	45.9	8719.9	90.8	71.4	62.2	53.7	32.6	12.4	18.3
MP6	8.5	1821.9	54.8	21330.3	75.1	61.8	74.3	64.5	51.7	49.3	28.1
NA1	0.8	8.3	36.5	1030.4	36.4	68.8	11.3	95.1	37.7	85.8	24.5
OR1	16.7	1369.7	63.0	8202.4	58.4	61.6	64.9	95.3	62.6	90.8	37.5
OR2	9.7	328.2	36.1	3399.3	90.4	90.2	86.4	86.6	33.3	63.6	31.5
OR3	16.0	787.8	39.0	4927.7	65.1	89.8	79.4	92.5	38.3	88.8	34.7
PO1	0.3	63.7	80.1	20927.5	27.6	59.6	52.6	85.6	82.0	85.6	46.0
PU1	9.9	2922.9	61.8	29451.3	64.5	50.2	35.8	83.2	58.6	33.0	25.1
PU2	8.5	4744.4	69.7	55704.8	68.1	45.9	68.1	65.9	61.5	26.7	28.2
RA1	15.3	3607.4	53.6	23523.2	62.1	33.1	80.1	28.3	54.3	14.2	9.4
RA2	20.8	4334.1	54.5	20884.0	55.2	29.9	78.3	77.2	53.7	58.6	10.9
RA3	10.2	406.3	37.6	3998.9	32.7	29.0	67.9	94.2	36.9	83.7	11.4
RA4	6.8	1404.2	65.3	20560.1	73.0	48.3	86.7	74.6	62.6	47.2	29.8

Contd...

NSS Regions	Far HH in Lakhs	Out Amt of Far HH in Rs Crore	Prop of Far HH Indebted, %	Avg Out Amt per Far HH in Rs	Prop of Out Amt for Prod purposes, %	Prop of Out Amt from Formal sources, %	Prop of Out Amt >1 Yr, %	Prop of SM Far HH to all Far HH, %	Prop of SM Far HH Indebted, %	Out Amt of SM Far HH to Out Amt of all Far HH, %	Out Amt of SM Far HH from Formal sources, %
SI1	0.5	10.9	32.8	2052.5	39.2	57.8	50.5	95.1	33.3	88.7	19.6
TN1	9.7	2680.3	72.9	27738.8	51.4	54.3	71.3	90.7	70.9	74.5	38.5
TN2	10.5	2562.9	80.3	24454.5	50.3	49.2	66.3	92.3	81.0	78.1	35.1
TN3	9.8	1577.1	74.4	16012.0	45.5	36.2	65.8	87.2	76.6	87.4	26.3
TN4	8.9	2496.4	69.3	28090.5	69.1	60.3	79.4	78.9	69.7	56.8	35.6
TR1	2.3	69.5	49.2	2977.2	59.1	79.7	89.8	99.5	49.4	100.0	39.4
UP1	53.4	6858.9	53.5	12834.6	72.1	61.9	63.8	89.7	52.9	69.7	27.0
UP2	33.8	1630.1	32.6	4820.0	66.5	62.3	71.6	91.2	32.0	75.7	18.9
UP3	75.1	3120.8	33.1	4154.2	55.8	52.8	70.5	93.7	32.6	75.9	13.6
UP4	9.2	1129.1	50.6	12284.4	77.5	63.4	69.7	69.4	44.0	35.2	19.7
UT1	9.0	99.3	7.2	1108.0	51.5	76.1	60.1	96.2	7.0	98.5	9.5
WB1	6.6	168.5	29.9	2566.5	74.7	94.6	77.6	97.0	30.5	98.9	28.8
WB2	19.8	790.1	40.3	3983.9	64.9	70.3	61.5	96.4	40.6	92.6	26.9
WB3	23.1	1664.1	63.2	7213.0	53.1	48.7	60.8	97.5	63.0	89.4	29.3
WB4	19.8	1002.8	50.9	5075.1	50.5	56.7	70.9	96.6	51.0	96.8	28.7
India	893.5	112445.3	48.6	12584.8	65.1	57.0	69.7	83.6	46.3	58.9	23.8

Note and Source: As in Appendix 3a. Region codes as in Appendix 2b.

Figure 1: Outstanding Amount per Indebted Farmer Household

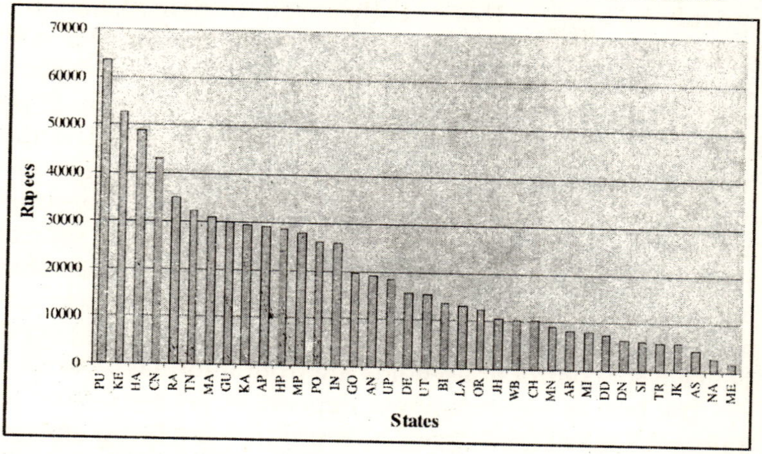

Note: AN=Andaman & Nicobal Islands, AP=Andhra Pradesh, AR=Arunachal Pradesh, AS=Assam, BI=Bihar, CH=Chattishgarh, CN=Chandigarh, DD=Daman & Diu, DE=Delhi, DN=Dadra & Nagar Haveli, GO=Goa, GU=Gujarat, HA=Haryana, HP=Himachal Pradesh, IN=India, JH=Jharkhand, JK=Jammu & Kashmir, KA=Karnataka, KE=Kerlal, LA=Lakshadweep, MA=Maharashtra, ME=Meghalaya, MI=Mizoram, MN=Manipur, MP=Madhya Pradesh, NA=Nagalan, OR=Orissa, PO=Pondicherry, PU=Punjab, RA=Rajasthan, SI=Sikkim, TN=Tamil Nadu, TR=Tripura, UP=Uttar Pradesh, UT=Uttaranchal, WB=West Bengal.

Source: Calculated from unit level data using 33rd schedule, 59th round of National Sample Survey (NSS) on Situation Assessment Survey of Farmers.

Source wise distribution suggests that 58 per cent are from formal sources (Government 2.5 per cent, Cooperatives 19.6 per cent and Banks 35.6 per cent); more than one-fourth of the outstanding debt is from moneylenders and the rest from other informal sources. The states with relatively higher proportion from moneylenders are Andhra Pradesh, Bihar, Manipur, Punjab, Rajasthan and Tamil Nadu. The proportion from moneylenders as well as other informal sources is much lower in Maharashtra where more than half of the loan is from Cooperatives and more than one-third from Banks. Nearly 70 per cent of the outstanding amount is from loans that are more than a year old, that is, they are carried over from earlier agricultural seasons. This is nearly three-fourths in Maharashtra with 84 per cent of these being from formal sources. When we consider the small and marginal farmers then the share of outstanding amount from formal sources is less than 25 per cent. This is of concern because 84 per cent of all farmer households are from this category. These are indicative of a huge unmet demand and as a consequence, a substantial amount of current loans is likely to be from informal sources. Field survey observations in western Vidarbha of Maharashtra indicates that from the outstanding debt of loans taken in 2004 nearly 72 per cent for suicide case households and about 38 per cent for control households are from informal sources.

The farmer is increasingly dependent on the market for inputs. Usage of seeds from own produce are being replaced by hybrid varieties and now with the costlier genetically modified varieties (particularly, in Cotton). Farmers are led into situations where they have to use greater and greater dosage of fertilizers to maintain productivity. Spraying of pesticides has increased over years as the pests develop immunity, but they also have an adverse effect by killing the friendly pests. With new inputs there is a change in technology of cultivation. The farmers existing knowledge becomes redundant; there is deskilling. In the absence of professional extension service, the farmers are dependent on advice from input dealers leading to supplier-induced-demand. There have been instances of villages opting for a second/third sowing during deficient rainfall years without any dependence on irrigation or even ground water. Besides overuse, usage of spurious varieties of inputs is also a matter of serious concern. Situations where seed sown have led to healthy growth of the plant but do not yield any produce lead to a suspicion that genetically modified terminator seeds are being sold under cover.

Case study of a farmer owning eight acres of unirrigated land in Yavatmal suggests the following. In 2004, he cultivated cotton in five acres where he had to go in for a second sowing due to delay in rain. This has led to an increase in seed expenses, but the expenses incurred in the second instance was reduced by half by going in for a different variety and using some left over seeds.[3] The total expenditure on seed was Rs. 7500/-. After including expenditure on fertilizer (Rs. 5000/-), pesticides (Rs. 3000/-) and labour (Rs. 2000/-) his total costs are Rs. 17500/-. He got a produce of 15 quintals, which he sold to the Maharashtra State Cooperative Cotton Growers Marketing Federation (MSCCGMF) through the monopoly cotton procurement scheme (MCPS). At the time of survey, he had received Rs. 1500/- per quintal and was expecting another Rs. 500/- per quintal. After receiving this balance amount, his net income will be Rs. 12500/-. The remaining three acres, used for cultivating crops for consumption purposes, under a deficient rain did not give much return. A good crop (say, 4 quintals of cotton per acre) would have taken this farmer above the poverty line, but now he is below the poverty line.[4] This depicts that the transient state of poverty of even a semi-medium farmer household. The situation would be worse for marginal/small farmers who are likely to have lower access to credit from formal sources. A tenant farmer will also have additional costs in the form of rent. Further, because of lower volumes of produce or immediate cash requirement or non-legal status of tenancy they may end up selling the produce to traders at lower then the price prevailing in market centres. A slight dip in the price of produce will also have an adverse effect on their income.

Opening up of the economy has led to certain cash crops like Cotton and Pepper among others being exposed to greater price volatility. Excess international supply of Cotton at a lower price is also because of direct and indirect subsidies leading to dumping by the USA. Domestic policies in India have led to removal of quantitative restrictions and subsequently reduction of import tariff from 35 per cent in 2001-02 to 5 per cent in 2002-03 increasing our vulnerability to the volatility of international prices. It is at this critical juncture when there is a greater need of price stabilization that the monopoly cotton procurement scheme (MCPS) of Maharashtra has become non-functional. Disbanding of this scheme in 2005-06 has in fact led to a reduction of Rs. 500/- per quintal advance additional price that has in recent years acted as a cushion against the higher costs in the state. The Commission for Agricultural Costs and Prices estimates the cost of production for cotton in Maharashtra at Rs. 2303/- per quintal, but the all India minimum support price for the long staple variety of fair average quality is only Rs. 1980/-.

The larger agrarian crisis has an adverse effect on farm households. A symptom of this crisis is reflected in the increasing incidence of farmers' suicides.

FARMERS' SUICIDES IN INDIA[5]

In recent years, farmers' suicides in India have attracted wide media, public policy and scholarly attention. In India, there have been 156,562 farmer suicides during 1995-2004. From these, more than four-fifths are males. The suicide mortality rate (SMR, suicide death per 100,000 persons) for male farmers nearly doubled in ten years from 9.7 in 1995 to 19.2 in 2004. SMR for male non-farmers has veered around 13; it increased from 12.6 in 1995 to 14.2 in 1999 and then decreased to 13.4 in 2004 (Figure 2).[6] Conditions should be created such that the

Figure 2: SMR for Male Farmers and Male Non-farmers in India, 1995-2004

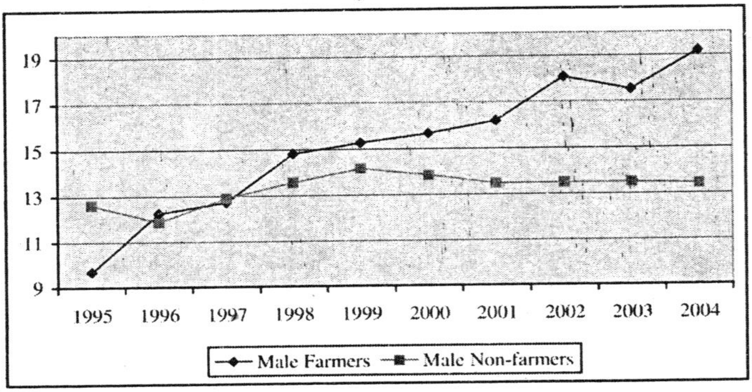

Note: SMR for farmers are based on interpolated/extrapolated population for cultivators using 1991 and 2001 census. For details, see Mishra (2006c).

Source: National Crime Records Bureau (NCRB) (Various Years).

possibilities of committing suicide should be reduced. Suicide being a rare event, relatively higher incidence among a sub-group could be indicative of a larger socio-economic malaise. For every individual committing suicide, there could be many more in a state of despair. Inclusive development should address the larger crisis.

In 2004, the states/union territories with SMR for male farmers higher than the national average are Andhra Pradesh (44.5), Goa (32.1), Karnataka (35.4), Kerala (183.0), undivided Madhya Pradesh (27.7) Maharashtra (57.2), Sikkim (40.5), Tamil Nadu (43.7), Dadra & Nagar Haveli (42.5), Delhi (49.4) and Pondicherry (1495.4).[7] Together these states account for nearly four-fifths, of farmers, suicide in 2004, more than half from the states of Andhra Pradesh, Karnataka, Kerala and Maharashtra, another one-fourth from undivided Madhya Pradesh and Tamil Nadu and the remaining five smaller states/union territories contributing for a little more than one per cent. As indicated earlier, higher farmers' suicide is symptomatic of a larger agrarian crisis. Some of the systemic risks (faced by a large number of households) have been discussed in the previous section.

Now, we take up discussion on idiosyncratic factors based on field survey in western Vidarbha, Maharashtra. From the 111 suicide cases analyzed, 91 per cent were males, 55 per cent in the age group of 31-50 years, and 80 per cent were currently married, nearly 40 per cent were matriculates or with higher education, 58 per cent had more than 10 years of experience. The size-class of land shows that 53 per cent were marginal and small farmers with less than 5 acres of land.

Analysis of suicide case household indicates that on average 4.8 risk factors were identified per case. Some of the socio-economic risk factors identified are indebtedness (87 per cent), deterioration of economic status (74 per cent, conflict with other members in the family (55 per cent), crop failure (41 per cent), decline in social position (36 per cent), burden of daughter's/sister's marriage (34 per cent), suicide in a nearby village (32 per cent), addictions (28 per cent), change in behaviour of deceased (26 per cent), dispute with neighbours/

others (23 per cent), health problem (21 per cent), a recent death in the family (10 per cent), history of suicide in the family (6 per cent) and other family members being ill (4 per cent). These factors are not mutually exclusive. They can co-exist and are also interrelated such that they can feed into each other and aggravate each other (see case studies in Box 1).

Box 1

Cases of Suicide Death by Farmers in Vidarbha

Case A: 45 years male. The household has 6 acres of land and has an annual income of Rs. 35000/-. Discussion revealed that: (i) delayed monsoon led to double/triple sowing increasing input costs for the household (ii) the deceased individual had entered into cotton trading for which he had taken loans and invested on other farmers and as a result the impact of crop failure by other farmers would also add to his burden, (iii) the individual had plans of getting at least one of his daughters married, (iv) he was also contemplating of contesting the local Sarpanch elections indicating that any economic downfall would effect his social reputation immensely. From this case one can infer that there is a complex interplay of multiple causes that are not mutually exclusive. After the demise of this individual there were three/four cases in nearby villages in about 7-10 days. This suggests a cascading effect. People from neighbouring villages who also had their own problems could relate to someone who is also a peasant like them.

Case B: A young man in his early 20s. Few years ago his father was not well and he took over the reigns of cultivation and experimented with input intensive cultivation. From the returns he could spend on father's treatment, expenses on a sister's marriage and improved the overall economic condition of the household. Other farmers also started taking advice from him. He was confident of a good crop and initiated plans for getting the other sister married. He had also told his mother that he would like to get married to a girl from a poorer household without taking any dowry. To his dismay, the crop failed.

Case C: 52 years, male. They had 52 acres of land at one time, but now they have 36 acres only–indicating a scenario where economic position is declining. The individual had outstanding loan from formal as well as informal sources and a couple of days before the fateful incident, the private moneylender had visited and insisted on being returned the money or transferring ownership of some land. There were two daughters of marriageable age. There was a burden on higher education of his children. The deceased had been talking some time ago with his children about other suicide cases and given his opinion that that is no solution; this perhaps gives a faint indication that he might have been contemplating suicide for some time.

Appendix 4a
Trends in Suicide Mortality Rates Across States, 1995-2004 (Farmer and Non-farmer Males)

States	*Farmers*										*Non-Farmers*									
States	*1995*	*1996*	*1997*	*1998*	*1999*	*2000*	*2001*	*2002*	*2003*	*2004*	*1995*	*1996*	*1997*	*1998*	*1999*	*2000*	*2001*	*2002*	*2003*	*2004*
Andaman & Nicobar Is	0.0	54.1	80.9	74.0	60.4	107.2	20.1	46.8	6.7	20.1	54.4	54.3	50.1	46.1	54.7	39.9	46.9	50.5	40.4	44.4
Andhra Pradesh	13.6	24.4	17.5	28.8	30.1	22.8	25.6	31.7	28.5	44.5	10.9	11.1	14.3	14.3	15.9	16.4	16.9	19.5	19.5	21.6
Arunachal Pradesh	0.0	0.0	8.5	0.0	2.3	5.2	8.8	15.2	7.2	11.3	11.0	11.7	9.9	10.2	12.2	17.6	17.1	13.0	11.1	8.1
Assam	4.5	2.7	7.5	5.6	2.5	4.2	5.5	9.6	6.1	12.6	14.0	19.3	21.9	20.4	18.3	19.7	17.6	14.7	15.0	15.6
Bihar*	1.0	0.7	0.9	1.1	1.1	0.3	0.9	1.0	0.8	0.5	1.3	1.2	1.8	2.3	2.2	0.5	0.9	1.2	1.0	0.9
Chandigarh	90.6	0.0	0.0	0.0	0.0	0.0	0.0	0.0	130.1	0.0	9.4	4.4	4.8	11.4	8.4	10.3	8.4	11.1	13.3	10.1
Dadra & Nagar Haveli	24.2	59.1	105.5	82.1	42.0	113.1	105.1	119.6	135.1	42.5	25.2	26.5	12.2	8.5	19.6	20.6	18.0	11.5	6.8	11.8
Daman & Diu	38.4	0.0	0.0	0.0	0.0	47.6	0.0	104.3	0.0	114.6	8.9	15.0	7.8	19.1	8.2	9.0	12.1	14.7	19.2	3.0
Delhi	37.1	70.1	32.9	125.4	66.4	61.3	59.6	115.5	31.8	49.4	10.7	9.6	10.0	8.3	9.0	9.1	10.4	8.6	9.9	10.2
Goa	39.8	77.8	22.1	16.6	17.5	55.3	70.0	37.0	73.8	32.1	23.1	58.4	22.6	29.2	28.6	28.5	24.5	35.1	31.1	32.0
Gujarat	9.6	11.3	11.0	14.0	10.3	14.9	13.2	12.4	12.1	11.7	10.2	10.3	8.8	10.4	12.1	11.9	10.8	11.0	10.5	11.2
Haryana	10.1	4.9	2.5	9.2	9.4	12.0	7.2	8.9	10.1	7.5	16.1	9.8	14.1	11.7	17.4	17.2	14.9	16.5	16.2	14.5
Himachal Pradesh	1.9	3.6	2.4	2.1	4.8	4.2	2.6	2.8	3.4	5.0	7.3	5.8	10.3	6.7	8.7	10.7	7.9	9.2	10.3	9.2
Jammu & Kashmir	0.0	0.0	0.3	0.2	0.1	3.0	1.2	1.5	0.5	0.1	0.8	0.5	1.1	1.3	1.4	1.2	2.0	2.0	1.4	1.2
Karnataka	33.7	30.9	31.3	30.1	41.4	43.5	44.5	41.6	48.2	35.4	31.3	23.2	27.3	30.0	31.3	30.5	29.3	30.4	29.6	30.2
Kerala	127.6	109.4	138.9	172.9	182.5	184.7	161.8	258.3	297.6	183.0	35.3	35.2	39.7	40.0	42.4	40.7	43.2	42.1	39.3	41.1
Lakshadweep	NA	NA	NA	NA	NA	NA	NA	NA	NA	NA	0.0	0.0	0.0	3.8	0.0	0.0	0.0	0.0	3.5	0.0
Maharashtra	14.7	23.5	23.9	29.0	30.6	37.3	44.1	47.3	50.8	57.2	18.0	14.6	16.5	17.0	16.3	16.5	16.5	15.8	15.7	14.5
Manipur	0.4	0.0	0.0	0.0	0.0	0.4	0.0	1.3	1.8	0.9	2.7	2.3	1.0	3.1	1.9	1.6	2.9	3.2	1.9	3.3
Meghalaya	3.7	3.7	0.4	2.0	1.2	0.8	0.8	3.1	1.5	4.2	5.2	7.9	8.7	7.3	4.8	4.0	8.9	5.2	3.5	3.3

Contd...

States	Farmers										Non-Farmers									
States	*1995*	*1996*	*1997*	*1998*	*1999*	*2000*	*2001*	*2002*	*2003*	*2004*	*1995*	*1996*	*1997*	*1998*	*1999*	*2000*	*2001*	*2002*	*2003*	*2004*
Mizoram	0.0	0.0	4.1	0.0	0.0	2.3	0.0	2.3	0.0	1.5	11.7	13.1	8.6	26.9	8.8	15.2	16.6	18.5	13.9	16.7
Madhya Pradesh*	10.5	14.4	17.9	18.4	21.8	22.1	24.3	22.5	21.2	27.7	12.1	9.6	10.6	14.4	13.3	15.6	14.2	14.5	14.3	13.5
Nagaland	1.9	1.8	3.5	0.0	0.4	0.0	0.0	0.0	0.0	0.0	3.1	4.3	7.1	1.0	1.1	4.2	3.7	2.6	2.2	3.1
Orissa	7.9	6.4	6.0	9.7	7.3	5.2	6.2	9.5	10.3	9.8	10.9	10.6	12.9	11.5	13.3	15.4	14.0	16.7	17.0	15.3
Pondicherry	0.0	0.0	291.8	1068.2	1145.1	2066.3	861.3	917.5	247.4	1495.4	108.6	81.5	75.1	53.2	67.2	36.9	55.2	73.8	81.1	47.4
Punjab	5.2	7.3	6.1	6.0	4.8	4.1	2.4	2.3	1.4	4.3	4.2	3.4	3.6	5.2	7.0	5.8	3.5	3.5	4.5	4.1
Rajasthan	0.0	6.0	8.3	8.9	8.4	7.6	6.0	6.9	7.9	9.4	13.1	7.6	7.7	8.6	9.6	8.1	8.3	8.3	9.2	8.9
Sikkim	0.0	11.0	24.7	21.2	22.2	11.6	22.8	7.0	27.5	40.5	25.3	29.4	10.9	29.7	32.0	21.1	24.0	24.7	23.3	12.2
Tamil Nadu	NA	15.5	20.2	26.4	18.3	20.9	25.3	39.7	27.1	43.7	NA	19.9	19.2	22.4	24.4	24.2	24.7	22.1	24.0	24.8
Tripura	45.0	48.2	7.5	50.8	42.2	6.2	18.6	12.5	0.0	7.2	20.4	16.2	29.8	21.2	35.4	44.7	33.2	35.8	36.6	30.6
Uttar Pradesh*	1.9	2.6	2.4	3.2	3.9	3.2	3.3	2.8	2.1	2.4	3.6	3.5	3.6	3.2	3.6	3.2	2.2	3.1	2.8	2.4
West Bengal	18.5	26.4	25.3	20.8	18.8	21.2	20.1	25.1	20.7	18.4	20.8	21.1	20.8	21.4	21.0	19.5	19.9	17.7	18.9	19.5
India	9.3	12.3	12.7	14.8	15.3	15.7	16.2	18.1	17.5	19.2	11.7	11.9	12.9	13.6	14.2	13.9	13.5	13.5	13.5	13.4

Note: Estimates for 1995 is not given for Tamil Nadu as appropriate profession wise classification is not available. * Undivided states.
Source: Estimated using interpolated/extrapolated population using census 1991 and 2001 and suicide data from NCRB (Various Years).

Appendix 4b

Trends in Suicide Mortality Rates Across States, 1995-2004 (Farmer and Non-farmer Females)

States	*Farmers*										*Non-Farmers*									
States	*1995*	*1996*	*1997*	*1998*	*1999*	*2000*	*2001*	*2002*	*2003*	*2004*	*1995*	*1996*	*1997*	*1998*	*1999*	*2000*	*2001*	*2002*	*2003*	*2004*
Andaman & Nicobar Is	0.0	0.0	0.0	0.0	0.0	0.0	0.0	0.0	0.0	0.0	42.8	53.0	42.5	42.8	29.4	35.8	35.4	36.3	28.6	26.4
Andhra Pradesh	13.6	11.5	4.6	8.7	13.0	11.8	6.7	10.4	14.4	18.7	9.6	10.0	11.9	12.1	12.6	11.4	12.5	11.9	10.7	11.9
Arunachal Pradesh	0.0	0.0	0.7	0.0	2.1	1.4	3.5	4.2	0.7	2.7	9.0	5.1	9.5	5.8	8.8	13.9	9.8	11.6	7.9	7.6
Assam	0.5	0.0	1.0	0.6	1.2	1.1	2.1	2.2	3.1	1.6	7.2	9.1	10.0	9.1	8.2	8.4	7.7	7.4	8.2	7.9
Bihar*	0.6	0.4	0.4	0.8	0.8	0.0	0.2	0.4	0.0	0.0	1.2	1.1	1.5	2.1	2.0	0.6	0.9	0.9	0.8	1.7
Chandigarh	0.0	0.0	0.0	0.0	0.0	0.0	0.0	0.0	0.0	0.0	4.9	3.7	9.4	12.2	10.2	8.4	8.7	9.2	9.1	5.8
Dadra & Nagar Haveli	0.0	23.9	28.7	8.4	0.0	0.0	0.0	0.0	0.0	0.0	14.6	11.3	31.4	23.4	16.4	27.2	23.9	26.9	30.9	25.1
Daman & Diu	0.0	0.0	0.0	0.0	0.0	0.0	0.0	0.0	0.0	0.0	13.2	8.5	8.1	11.8	7.6	7.3	7.1	3.4	10.0	12.9
Delhi	10.3	50.8	0.0	0.0	0.0	0.0	18.9	28.0	0.0	9.1	7.9	8.7	8.6	9.0	8.6	7.7	9.1	7.0	6.8	7.2
Goa	5.8	0.0	0.0	0.0	0.0	0.0	0.0	8.6	4.6	4.8	19.8	37.2	15.2	15.0	14.7	15.1	15.2	13.6	14.2	16.4
Gujarat	5.7	3.9	5.5	4.5	4.5	3.8	4.1	4.9	6.7	4.5	10.9	11.0	10.4	10.9	11.9	11.0	10.4	9.0	8.6	8.6
Haryana	0.0	0.0	0.0	2.1	3.4	1.7	1.0	1.9	1.1	1.1	9.1	5.5	7.6	7.6	10.2	8.3	8.3	7.8	7.6	7.6
Himachal Pradesh	0.0	0.2	0.4	0.9	0.0	0.0	0.0	0.1	0.4	0.7	6.8	5.3	7.2	6.2	7.9	7.4	8.1	7.7	8.6	7.6
Jammu & Kashmir	0.0	0.0	0.0	0.0	0.4	1.4	0.5	0.2	0.5	0.0	0.6	0.7	1.7	1.7	1.3	1.6	1.6	2.3	1.8	1.5
Karnataka	35.3	21.7	13.9	19.0	17.2	24.8	17.2	16.7	18.3	13.9	18.6	14.5	18.0	17.8	20.8	18.6	17.0	17.5	16.1	15.9
Kerala	21.2	17.1	23.4	97.9	84.1	62.4	50.0	95.7	70.5	56.3	16.9	18.8	19.0	18.2	19.2	17.7	18.3	16.9	16.0	15.6
Lakshadweep	NA	NA	NA	NA	NA	NA	NA	NA	NA	NA	0.0	0.0	0.0	0.0	0.0	0.0	0.0	0.0	3.7	0.0
Maharashtra	1.8	7.1	5.6	8.5	6.9	10.1	11.5	10.8	9.4	7.4	15.9	13.6	15.3	15.4	15.0	13.3	12.7	12.0	11.5	11.2
Manipur	0.0	0.0	0.0	0.0	0.0	0.0	0.0	0.0	0.0	0.0	0.8	0.3	0.8	0.9	1.1	1.6	2.0	1.0	0.5	0.9
Meghalaya	0.0	1.9	0.0	0.5	0.9	0.0	0.9	0.0	0.0	0.0	2.6	1.4	2.3	2.3	2.4	2.4	2.1	2.2	1.1	1.9

Contd...

States	*Farmers*										*Non-Farmers*									
States	*1995*	*1996*	*1997*	*1998*	*1999*	*2000*	*2001*	*2002*	*2003*	*2004*	*1995*	*1996*	*1997*	*1998*	*1999*	*2000*	*2001*	*2002*	*2003*	*2004*
Mizoram	0.0	0.0	0.0	0.0	0.0	0.0	0.0	0.0	0.0	0.0	2.0	1.4	1.8	8.9	2.1	2.9	3.2	3.8	4.0	2.5
Madhya Pradesh*	3.7	7.0	11.0	8.6	9.7	9.6	9.1	8.2	9.5	8.3	12.0	11.1	10.3	13.6	13.5	14.7	14.6	14.2	13.2	34.7
Nagaland	0.0	0.4	0.0	0.0	0.0	0.0	0.0	0.0	0.0	0.0	0.8	1.7	2.0	0.8	1.4	0.9	2.6	1.3	0.8	0.9
Orissa	4.0	1.4	2.4	5.8	0.6	2.0	5.4	4.3	5.3	11.2	11.0	10.6	10.0	11.4	11.4	11.3	12.4	10.9	10.1	9.6
Pondicherry	0.0	0.0	427.7	199.4	465.1	1220.9	633.4	462.4	216.1	3285.5	48.0	50.2	44.7	44.4	51.9	41.8	44.6	33.5	41.4	36.9
Punjab	1.4	0.4	0.0	0.0	0.0	0.0	1.0	0.0	0.3	0.3	1.9	1.7	1.8	2.3	2.4	1.9	2.5	1.1	1.2	1.2
Rajasthan	0.0	0.8	1.6	1.6	2.4	3.4	1.3	1.5	1.2	1.1	8.2	6.4	8.1	7.0	8.1	6.4	6.6	5.9	6.4	6.1
Sikkim	0.0	2.0	11.3	1.8	3.5	3.4	3.3	4.7	16.8	28.1	15.9	13.3	15.0	13.9	21.8	21.8	18.1	12.3	14.8	12.7
Tamil Nadu	NA	9.5	8.8	7.0	8.4	9.3	8.6	11.2	12.9	18.6	NA	14.6	14.8	17.6	18.0	15.4	15.0	15.0	16.3	16.4
Tripura	131.8	123.4	0.0	128.5	0.0	0.0	0.0	15.3	4.4	0.0	11.4	11.5	27.3	14.8	32.5	28.3	29.6	20.2	25.6	24.1
Uttar Pradesh*	0.4	4.0	2.8	2.7	2.5	2.7	2.0	0.9	0.7	1.4	3.1	3.3	3.0	3.9	4.0	3.2	2.6	3.0	2.5	5.3
West Bengal	24.9	30.0	21.7	37.3	29.7	35.3	31.1	39.2	12.7	2.9	18.7	20.3	20.2	19.5	18.6	18.6	17.3	15.9	16.2	15.9
India	5.6	6.6	5.6	7.2	6.7	7.4	6.2	6.4	5.9	5.5	8.9	9.6	10.2	10.7	11.1	10.1	9.9	9.3	9.1	9.0

Note: Estimates for 1995 is not given for Tamil Nadu as appropriate profession wise classification is not available. * Undivided states

Source: Estimated using interpolated/extrapolated population using census 1991 and 2001 and suicide data from NCRB (Various Years).

Appendix 4c

Trends in Suicide Mortality Rates Across States, 1995-2004 (Age-Adjusted Population for Males and Females)

States	*Males*										*Females*									
States	*1995*	*1996*	*1997*	*1998*	*1999*	*2000*	*2001*	*2002*	*2003*	*2004*	*1995*	*1996*	*1997*	*1998*	*1999*	*2000*	*2001*	*2002*	*2003*	*2004*
Andaman & Nicobar Is	49.1	54.3	53.0	48.7	55.2	45.7	44.6	50.2	37.7	42.5	40.2	50.0	40.2	40.5	27.9	34.1	33.8	34.8	27.5	25.4
Andhra Pradesh	11.4	13.3	14.8	16.6	18.1	17.4	18.2	21.2	20.7	24.7	10.0	10.1	11.2	11.8	12.7	11.5	12.0	11.8	11.0	12.3
Arunachal Pradesh	7.9	8.4	9.5	7.4	9.5	14.3	14.9	13.6	10.1	8.9	5.8	3.3	6.5	3.9	6.6	9.8	7.8	9.3	5.7	6.1
Assam	11.5	15.1	18.4	16.9	4.6	16.2	15.0	13.7	13.2	15.0	6.3	7.9	8.9	8.1	7.4	7.7	7.1	7.0	7.8	7.4
Bihar*	1.2	1.1	1.6	2.1	2.0	0.5	0.9	1.1	1.0	0.8	1.1	1.1	1.4	2.0	1.9	0.6	0.9	0.9	0.8	1.6
Chandigarh	9.9	4.4	4.7	11.3	8.3	10.3	8.4	11.0	13.6	10.1	4.8	3.7	9.3	12.1	10.2	8.4	8.7	9.2	9.1	5.8
Dadra & Nagar Haveli	24.9	34.7	34.0	24.4	24.1	37.6	32.8	28.3	25.1	15.8	9.0	15.8	30.5	18.7	11.5	19.7	17.8	20.6	24.2	20.1
Daman & Diu	10.2	14.4	7.5	18.4	8.0	10.0	11.8	16.7	18.8	4.9	12.1	7.9	7.6	11.1	7.2	7.0	6.8	3.3	9.7	12.6
Delhi	10.9	9.9	10.1	8.9	9.3	9.3	10.6	9.0	9.9	10.3	7.9	8.8	8.6	8.9	8.6	7.7	9.1	7.0	6.8	7.2
Goa	24.2	59.5	22.6	28.5	28.1	29.6	26.4	35.1	32.6	32.0	18.9	35.0	14.4	14.2	14.0	14.5	14.6	13.4	13.9	16.0
Gujarat	10.1	10.5	9.2	11.1	11.8	12.4	11.2	11.2	10.7	11.3	10.2	10.1	9.8	10.2	11.2	10.3	9.8	8.7	8.4	8.3
Haryana	14.8	8.8	11.8	11.2	15.9	16.3	13.5	15.1	15.1	13.3	8.1	4.9	6.7	7.0	9.4	7.4	7.3	7.0	6.6	6.6
Himachal Pradesh	5.7	5.1	7.8	5.3	7.5	8.7	6.3	7.3	8.3	8.0	4.2	3.3	4.5	4.1	4.7	4.4	4.8	4.5	5.2	4.7
Jammu & Kashmir	0.6	0.4	0.9	1.1	1.1	1.6	1.9	1.9	1.2	1.0	0.5	0.6	1.5	1.5	1.1	1.6	1.4	2.0	1.6	1.3
Karnataka	31.8	24.8	28.1	30.0	33.4	33.2	32.3	32.6	33.2	31.2	20.5	15.3	17.6	18.0	20.4	19.2	17.0	17.5	16.3	15.7
Kerala	42.0	40.2	45.7	47.4	49.6	47.4	48.3	50.5	48.5	45.8	17.0	18.8	19.1	19.3	19.9	18.1	18.5	17.4	16.3	15.8
Lakshadweep	0.0	0.0	0.0	3.8	0.0	0.0	0.0	0.0	3.5	0.0	0.0	0.0	0.0	0.0	0.0	0.0	0.0	0.0	3.7	0.0
Maharashtra	17.4	16.0	17.7	18.9	18.5	19.6	20.6	20.3	20.6	20.3	13.6	12.6	13.8	14.4	13.9	12.9	12.6	11.9	11.3	10.8
Manipur	2.1	1.7	0.7	2.3	1.5	1.4	2.3	2.8	1.9	2.8	0.6	0.2	0.6	0.7	0.9	1.3	1.7	0.9	0.5	0.8
Meghalaya	4.8	6.7	6.4	5.9	3.9	3.1	6.8	4.7	3.0	3.5	1.9	1.5	1.7	1.9	2.1	1.9	1.8	1.8	0.9	1.5

Contd...

States	Males										Females									
States	1995	1996	1997	1998	1999	2000	2001	2002	2003	2004	1995	1996	1997	1998	1999	2000	2001	2002	2003	2004
Mizoram	7.8	8.7	7.1	18.0	5.9	11.1	11.3	13.4	9.5	12.0	1.2	0.9	1.2	5.7	1.4	1.9	2.1	2.6	2.7	1.7
Madhya Pradesh*	11.6	10.9	12.6	15.5	15.6	17.3	16.7	16.5	15.9	16.80	10.2	10.2	10.4	12.6	12.8	13.7	13.7	13.2	12.6	30.6
Nagaland	2.7	3.5	6.1	0.7	0.9	3.0	2.6	1.9	1.6	2.3	0.5	1.2	1.3	0.6	0.9	0.6	1.8	0.9	0.5	0.6
Orissa	10.1	9.6	11.3	11.1	12.0	13.3	12.4	15.4	15.7	14.4	10.3	9.8	9.4	11.0	10.7	10.8	12.0	10.5	9.9	9.7
Pondicherry	104.8	78.9	81.6	81.2	94.7	84.7	72.7	90.6	84.2	72.0	47.6	49.8	46.9	45.2	53.6	45.8	46.2	34.5	41.7	42.3
Punjab	4.4	4.1	4.0	5.3	6.6	5.5	3.4	3.3	4.1	4.1	1.9	1.7	1.8	2.2	2.4	1.9	2.4	1.0	1.1	1.2
Rajasthan	9.1	7.1	7.9	8.7	9.3	8.0	7.6	8.0	8.9	9.0	6.0	4.9	6.4	5.6	6.6	5.6	5.2	4.8	5.1	4.8
Sikkim	18.0	24.2	14.7	27.3	29.3	18.5	23.7	20.0	24.4	19.5	11.7	10.3	14.0	10.7	16.9	16.8	14.1	10.3	15.3	16.8
Tamil Nadu	19.6.0	19.3	19.3	22.9	23.6	23.8	24.8	24.0	24.3	26.7	13.2.0	14.1	14.4	16.8	17.3	14.9	14.6	14.7	16.1	16.5
Tripura	24.9	21.9	26.0	26.1	36.5	38.8	31.0	32.5	31.6	27.5	20.6	19.9	25.3	22.9	30.2	26.4	27.7	19.9	24.3	22.6
Uttar Pradesh*	3.1	3.3	3.3	3.2	3.6	3.2	2.5	3.1	2.7	2.4	2.9	3.3	3.0	3.8	3.9	3.1	2.5	2.8	2.4	5.0
West Bengal	20.4	21.9	21.5	21.3	20.7	19.7	19.9	18.6	19.1	19.3	19.0	20.6	20.3	20.1	18.9	19.1	17.7	16.5	16.1	15.5
India	12.5	11.9	12.9	13.8	14.4	14.2	14.0	14.3	14.2	14.4	9.5.6	9.3	9.7	10.4	10.6	9.8	9.5	9.1	8.8	8.7

Note: Age-adjusted population refers to 5 + years as suicide is not medically defined for 0-4 age-group. * Undivided states.Source: Estimated using interpolated/extrapolated population using census 1991 and 2001 and suicide data from NCRB (Various Years).

the National Sample Survey (NSS) Coastal region and the other in the NSS Inland Eastern region), indicates that nearly half the below poverty line households were in a situation where all family members did not get two square meals a day at some time or the other during the year (Table 5). This survey was designed to choose 80 per cent poor households in the sample in areas that might be characterized as less than average developed. Adjusting for this, one gets roughly 6-8 per cent rural households facing food shortages at some time or the other during the year. This figure is in sharp contrast to hunger incidence of 3 per 1,000 rural households or 5 per farmer household in the state reported by National Sample Survey Organisation (NSSO) data for 2003 (59th round).[12]

Table 5
Food Deficit below Poverty Line Households across Seasons (%)

Season	*Jawhar, Thane*	*Yavatmal*	*Toal*
N	88	81	169
Winter	14	11	12
Summer	25	19	22
Monsoon	51	31	41
Any one season	48	17	33
Any Two seasons	6	9	7
All three seasons	10	9	9
At some time	64	35	50

Note: N indicates number of below poverty line households surveyed.
Source: Panda and Mishra (2005).

Across seasons, vulnerability was higher during the monsoon months. Many of the food insecure households resorted to migration to make both ends meet. This also affected their utilization of benefits from public facilities like Anganwadi and schools that existed in their villages. Their vulnerability can be understood from the case of a child death discussed in Box 2. One also observed non-payment of wages under public works and denial of food subsidies by not providing appropriate ration cards. There were also instances of some success stories-The "Wadi Project" (horticulture development) linked with MEGS and other programmes lead to improved livelihood opportunities (see Box 3) because these

Box 2

Case of Child Death in Jawhar

The anganwadi centre of Nagarmoda village reported 3 deaths and 1 stillbirth in the last two years. One case is from a below poverty line household which also faced food shortage at times. The family loses a son who is a little more than three years of age in November 2003. The boy who suffered from respiratory ailment was taken to the Rural Hospital at Jawhar. After initial treatment the doctor recommended that he be taken to JJ Hospital in Mumbai. Medical expenses would have been taken care of at the public health facility, but other opportunity costs made the family decide to return. Back in the village, neighbours, the anganwadi worker and gram sevak coaxed the family members to take the child to JJ Hospital. A few days later, after trying their luck with local healers, they returned to the public health facility at Jawhar, but it was too late. For this case, the family had sought treatment from local healers, public and private facilities and spent Rs.5000 on medical expenses and Rs.1000 on travel. The family now has three surviving children; one is the deceased boy's twin sister.

Box 3

Successful Utilization of Various Schemes by BAIF in Jawhar

Bharat Agro Industries Foundation (BAIF), a non-governmental organization (NGO), operates in tribal regions of 8-9 states of India. They have some presence in Jawhar taluka of Thane distrtict. Interventions by BAIF in some villages are very recent. We happened to visit a successful experiment in the village of Kelichapada where BAIF has been present for 6-7 years. One of their major interventions is through development of Wadi's, which when literally translated means orchard. To begin with, self-help groups are formed with 6-8 beneficiaries, each having 2-3 acres of land adjacent to each other. In the initial years the beneficiaries in Kelichapada worked on (a) their own plots leading to land development and planting of horticultural crops like cashew, guava and lemon and (b) construction of a water harvesting structure on a nearby stream. The inputs/help provided by BAIF came from different schemes. The formation of self-help groups under Swarnjayanthi Gram Swarozgar Yojana (SGSY), wages for their labour from Sampoorna Grameen Rozgar Yojana (SGRY)/ Maharashtra Employment Guarantee Scheme (MEGS) for three years – the time required for trees to bear fruit, non-wage expenses for land development, water harvesting structure and sapling of fruit bearing trees from MEGS or other government schemes. BAIF has also helped them organize and sell their produce in the market. Today the households no more migrate. The enrolment, retention and attendance of school going children has improved, their consumption and nutritional intake has increased and annual household incomes have more than doubled. BAIF's intervention has also helped landless households take up livestock rearing or other activities.

recent years also coincided with a spate of farmer suicides which was particularly high in the Inland Eastern (western Vidarbha) region of Maharashtra.

The problems of NREGS in a recent survey in Jharkhand point out the following difficulties: selling of application form for job cards, issuing of job cards to only below poverty line households and providing only one card for joint (not nuclear) households, providing defective job cards with no space to record wage payments, not maintaining muster rolls at worksites, fudging muster rolls, flawed work measurement, non payment of minimum wages, delayed wage payment. People are not aware that they have to apply for work and get a dated receipt for the same, unavailability of worksite facilities like shade for rest, drinking water and crèche, creation of productive assets is not up to the mark, fictitious gram sabhas, absence of elected representatives at local levels as elections to local bodies have not been held, involvement of middlemen, and excessive work load on local level government functionaries are some more of the problems. A public meeting held at the end of the survey was attended by 2000 poor labourers from neighbouring villages where pointed questions to the Block Development Officer (BDO) and gram sevaks and the local elected member of the legislative assembly who initially wanted to scuttle the meeting but having failed to do so he had to turn up. This passed on the message that local officials and elected representative are ultimately accountable to the people (Bhatia and Dreze 2006).[13]

One of the major limitations of the current employment guarantee is to restrict employment guarantee to only 100 days per household. This restriction is with the understanding that these households get some employment from other avenues, particularly agriculture. If agriculture is under strain then this assumption will not hold. Thus, in years of agrarian distress or in relatively poorer regions or for some other reason if households demand employment beyond 100 days then the administrative mechanism should try to address it. A failure in this regard will not come under the legal scanner of entitlement under a guarantee, but it would certainly be a

Table 6
Item-wise Share of MEGS Expenditure Across NSS Regions of Maharashtra, 2000-01 to 2003-04 (%)

Year/Item	*Coastal*	*Inland Western*	*Inland Northern*	*Inland Central*	*Inland Eastern*	*Eastern*	*Maharashtra**	
Rural population, 2001	11.1	28.3	14.4	21.1	17.3	7.8	100.0	(5.6)
Rural poor, 1999-2000	8.8	13.7	18.8	21.3	23.2	14.2	100.0	(1.3)
Item								
Roads#	11.5	12.8	14.4	38.0	13.0	10.3	100.0	(954.8)
Agriculture#	8.4	36.3	6.1	36.7	4.2	8.3	100.0	(929.9)
Irrigation#	0.3	17.2	7.7	53.4	6.5	15.0	100.0	(528.6)
Forestry#	10.7	10.6	13.7	33.2	17.9	14.0	100.0	(331.1)
Jawahar Wells#	7.0	20.5	14.4	25.7	21.6	10.7	100.0	(218.1)
Horticulture#	40.9	19.8	7.4	16.1	13.0	2.8	100.0	(211.9)
Establishment#	2.6	19.2	12.2	36.5	13.5	16.0	100.0	(86.7)
Miscellaneous#	5.3	20.4	19.1	43.1	5.9	6.3	100.0	(76.9)
Total#	10.0	21.1	10.6	37.5	10.4	10.5	100.0	(3338.0)
Total, 10 years$	10.4	21.5	9.8	35.8	11.4	11.1	100.0	(5523.4)

Note: * Figures in parentheses indicate total (in crore: number of population/poor and expenditure in rupees). For expenditure under MEGS it excludes certain miscellaneous expenditure at the aggregate level for the state. # Item-wise as well as total expenditure has been combined for four years: 2000-01 to 2003-04. $ Total 10 years data are average for 1994-95 to 2003-04.

Source: Mishra and Panda (2006).

failure in the domain of welfare. The success of the programme will also require that the provision is widely disseminated and people are made to be involved in at various stages. A mass social audit undertaken in Dungarpur,[14] a relatively poorer tribal district of Rajasthan from where seasonal migration is widespread, suggests that with a supportive administration and people willing to monitor the ongoing scheme can contain the corruption to a large extent and create assets that not only reflect people's demand (in this drought-prone region first priority was given to building/repairing water-harvesting structures) but would also help in enhancing livelihood opportunities (Sivakumar 2006).

CONCLUDING REMARKS

The situation is very complex. As per the 2003 situation assessment survey of farmers, 40 per cent of the farmers do not want to continue in the profession. They foresee a life of poverty and they would like to get out of this. The surge in farmers' suicides, which is symptomatic of a larger agrarian crisis, seems to be spreading. Relief measures should address all possible risks, the most important being the inadequate income that agriculture provides for the large mass of farmers. The income shortfall is further accentuated because of crop loss, market uncertainties (increasing input costs and decline in output prices) and additional expenditure requirements (health needs, ward's education and daughter's marriage among others). Without adequate safeguards, the farmer will require more and more credit that will lead him to a quagmire of indebtedness. Interventions in the credit market are required (particularly, to reduce the high cost of borrowing from informal sources), but on their own they are not likely to achieve much. Policy interventions should independently address all possible risks: income shortfalls, crop loss (weather, pests, theft, fire or spurious quality of seeds and other inputs), price shocks, increasing input costs and other uncertainties. Social safety measures should look into health, education and other relevant expenses. If farmers need a strong state support to enhance their livelihoods then the mass of agricultural labourers who depend on farmers would require even greater support.

Additional employment opportunities might provide some succour. Public interventions in employment have tried to focus on an appropriate wage rate (a balance between providing decent income and also one that would lead to targeting through self-selection by the poor), the nature of wage payment (some payments in kind to ensure food security and some in cash to meet non-food requirements) and in generating productive assets for the rural community. Some criticisms of these programmes are that the number of days of employment available was low, wage payments were delayed, kind payments were of poor quality, unavailability of work site facilities, creation of productive assets is not up to the mark and most important is siphoning of funds through corruption among others. A major channel of leakage was through fudging of muster roles. The provisions of transparency and public monitoring at each and every stage of EGS work (planning, implementation and stock-taking) under NREGA allows for evaluation of on-going work through social audit. This would check corruption much more than the annual checks which can only unearth book-keeping anomalies. Despite the limitation of restricting employment guarantee to only 100 days per household, "NREGA has created a sense of hope amongst the rural poor. This sense of hope can be further strengthened if people understand that the act gives them employment as a matter of right, and that claiming this right is within the realm of possibility" (Bhatia and Dreze 2006).

To sum up, the post-reform agrarian scenario has been a story of distress, despair and death. This, however, is not the end of the tunnel. For every death there are thousands in distress who continue to struggle and hope for a better tomorrow. In tackling agrarian crisis or ensuring employment guarantee one has to build on this hope.

NOTES

1. For some recent discussions on the agrarian crisis see Rao and Gopalappa (2004), Mohanty (2005), Reddy (2006), Reddy and Mishra (2006), Vaidyanathan (2006) and Vyas (2004) among others.

2. Value of output in agriculture in constant 1993-94 prices was lower in 2002-03 in comparison with the previous year by about 12 per cent. To account for this if one increases the returns from cultivation by one-third then also it would be less than Rs. 15,000/- per annum. Given a family size of 5.5 the per capita per day returns from cultivation turn out to be less than Rs. 8/-. Under such scenario, other sources become necessary if the farmer household has to stay above the poverty line.
3. It is generally the case that in an acre of land one packet of seeds (910 grams) that costs around Rs. 450/- to Rs. 500/- for non-Bt varieties and Rs. 1600/- for legal Bt varieties would suffice (in 2006-07 agricultural season, due to a court judgement price of legal Bt varieties have come down to about Rs. 1250/- per packet). However, due to a guaranteed germination rate of 65 per cent only, farmers end up sowing two instead of one seed and thereby increasing the seed requirement. Under assured water, such practices might reduce.
4. Updating the Planning Commission poverty line for rural Maharashtra to 2004 one gets an income of Rs. 4037/- per person per annum (that is, Rs. 336.45 per capita per month).
5. Some related discussions are in Mishra (2006a, 2006b, 2006c, 2006d and 2006e).
6. The SMR estimates of 2004 are based on provisional data. On the method used for calculating SMRs see Mishra (2006c). For trends in SMR for farmers, non-farmers and age-adjusted general population from 1995 to 2004 see Appendix 4.
7. The high SMR among farmers in Pondicherry is because of their low population, less then 10000 were cultivators as per 2001 census.
8. In the step-wise procedure, a variable is added if it increases chi-square significance by 0.05 and it is dropped if it increases chi-square significance by 0.1.
9. For an elaborate discussion on footloose labour see Breman (1996).
10. See, for example, Planning Commission (2000), Policy and Development Initiatives (2000), and Sen (2003) among others.
11. The situation was more severe in Jawhar, a tribal *taluka* in Thane (NSS Coastal Region), which is about three/four hours drive from Mumbai.

12. At the all India level, 59th round estimates indicate that 16 per 1000 rural households and 12 per 1000 farmer households were food deficit households. Across states, the maximum number of food deficit households is in Orissa: 71 per 1000 rural households and 54 per 1000 farmer households. The estimates for rural households and farmer households are based on different sample frames.
13. Another positive aspect indicating how different perspectives can come together is that the survey has been conducted by students from Delhi School of Economics and Jawaharlal Nehru University.
14. This social audit involved activists and researchers from across the country and also a few from outside India.

REFERENCES

Barrett, C., S. Holden and D. Clay (2004), Can Food-for-Work Programmes Reduce Vulnerability, Discussion Paper, Agricultural University of Norway.

Bhatia, B. and J. Dreze (2006), Employment Guarantee in Jharkhand: Ground Realities, *Economic and Political Weekly*, Vol. 41, No. 29, pp. 3198-3202.

Breman, J. (1996), *Footloose Labour: Working in India's Informal Economy*, Cambridge University Press, Cambridge.

Dandekar, A., S. Narawades, R. Rathod, R. Ingle, V. Kulkarni and Y. D. Sateppa (2005), *Causes of Farmer Suicides in Maharashtra: An Enquiry*, Final Report Submitted to the Mumbai High Court, Tata Institute of Social Sciences, Rural Campus, Tuljapur.

Mishra, S. (2006a), *Suicide of Farmers in Maharashtra*, Report submitted to the Government of Mahrashtra, Indira Gandhi Institute of Development Research, Mumbai.

Mishra, S. (2006b), Farmers' Suicides in Maharashtra, *Economic and Political Weekly*, Vol. 41, No. 16, pp. 1538-1545.

Mishra, S. (2006c), Suicide Mortality Rates across States of India, 1975-2001: A Statistical Note, *Economic and Political Weekly*, Vol. 41, No. 16, pp. 1566-1569.

Mishra, S. (2006d), Farmers' Suicides in Vidarbha, Mahrashtra, in Alternative Survey Group *Alternative Economic Survey, India 2005-06: Disempowering Masses*, Danish Books, Delhi, pp. 69-71.

Mishra, S. (2006e), Suicides in India: Some Observations, in K. S. Bhat and S. Vijaya Kumar (eds.) *Undeserved Death: A Study on Suicide of Farmers in Andhra Pradesh (2000-2005)*, Council for Social Development, Hyderabad, pp. 93-113.

Mishra, S. and M. Panda (2006), Growth and Poverty in Maharashtra, WP-2006-001, Indira Gandhi Institute of Development Research, Mumbai.

Mohanty, B. B. (2005), 'We are Like the Living Dead': Farmer Suicide in Maharashtra, Western India, *Journal of Peasant Studies*, 32 (2): 243-76.

NCRB (Various Years), *Accidental Deaths and Suicides in India*, Ministry of Home Affairs, New Delhi.

NSSO (2005), *Situation Assessment Survey of Farmers: Indebtedness of Farmer Households*, NSS 59th Round (January-December 2003), Report No. 498 (59/33/1).

NSSO (2006), *Livestock Ownership Across Operational Landholding Classes in India, 2002-023*, NSS 59th Round (January-December 2003), Report No. 493 (59/18.1/1).

Panda, M., S. Mishra, S. Kamdar and M. Tondare (2005), *Evaluation of Food-for-Work (FFW) Component of Sampoorna Grameen Rozgar Yojana (SGRY) in Selected Districts of Maharashtra*, Report Submitted to Planning Commission, Indira Gandhi Institute of Development Research, Mumbai.

Panda, M. and S. Mishra (2005), *Poverty Reduction Strategy as Implementation of the Right to Development in Mahrashtra*, Report Submitted to Centre for Development and Human Rights, Indira Gandhi Institute of Development Research, Mumbai.

Planning Commission (2002), *Study on Employment Assurance Scheme (EAS)*, Programme Evaluation Organisation, *http://planning commission.nic.in/reports/peoreport/cmpdmpeo/volume1/178.pdf.*

Policy and Development Initiatives (2000) Study on Assessment of Poverty Alleviation Schemes in Maharashtra, Submitted to Planning Commission, *http://planningcommission.nic.in/reports/sereport/sereporf.htm.*

Rao, V. M. and D. V. Gopalappa (2004) Agricultural Growth and Farmers Distress: Tentative Perspectives from Karnataka, *Economic and Political Weekly*, Vol. 39, No. 52, pp. 5591-5597.

Ravallion, M., G. Datt and S. Chaudhuri (1993), Does Maharashtra's Employment Guarantee Scheme Guarantee Employment? Effects of the 1988 Wage Increase, *Economic Development and Cultural Change*, Vol. 41, No. 2, pp. 251-275.

RBI (2005), Handbook of Statistics on Indian Economy, 2005, RBI, Mumbai.

Reddy, D. N. (2006), Economic Reforms, Agrarian Crisis and Rural Stress in India, Mimeo, Hyderabad.

Reddy, D. N. and S. Mishra (2006), Crisis in Agriculture and Rural Distress in Post-reform India, prepared for K. Parikh and R. Radhakrishna (eds.) *India Development Report*, Oxford University Press, forthcoming.

Sen, S. (2003), Food and Employment Schemes in Barabanki, UP, http://*www.geocities.com/righttofood/data/upreport.pdf.*

Shetty, S. L. (2006), Monetary Policy and Financial Sector Liberalization, Chapter in *Macroeconomics of Poverty Reduction: India Case Study*, Report submitted to United Nations Development Programme, Indira Gandhi Institute of Development Research, Mumbai, April.

Singh, S. P. (2006), Agriculture, in Alternative Survey Group *Alternative Economic Survey, India 2005-06: Disempowering Masses*, Danish Books, Delhi, pp. 45-68.

Sivakumar, S. K. (2006), Working with a Purpose, *Frontline*, Vol. 23, No. 9, May 6-19.

Vaidyanathan, A. (2006), Farmers' Suicides and the Agrarian Crisis, *Economic and Political Weekly*, Vol. 41, No. 38, pp. 4009-4013.

Vakulabharanam, V. (2005), Growth and Distress in a South Indian Peasant Economy During the Era of Economic Liberalisation, *Journal of Development Studies*, Vol. 41, No. 6, pp. 971-997.

Vatsa, K. S. (2005), *Asset Based Approach to Household Risk Management: An Analysis of Selected Case Studies in South Asia*, Ph.D Dissertation, School of Engineering and Applied Science, George Washington University.

Vyas, V. S. (2004), Agrarian Distress: Strategies to Protect Vulnerable Sections, *Economic and Political Weekly*, Vol. 39, No. 52, pp. 5576-5581.

2

AGRICULTURAL EMPLOYMENT IN INDIA IN POST-REFORM PERIOD

NARAYAN C. NAYAK & FALGUNI PATTANAIK

INTRODUCTION

Historically, agricultural development is considered as the cornerstone of economic development. Development economists in general and agricultural economists in particular have long focused on how agriculture can best contribute to economic growth and modernization. Many early economists (Rosenstein-Rodan, 1943; Lewis, 1954; Scitovsky, 1954; Hirschman, 1958; Jorgenson, 1961) have treated agriculture pivotal because of its abundance and its ability to transfer surpluses to the industrial sector. Agriculture has the potential to create both production and consumption linkages with the non-farm sector (Mellor, 1976).

As India has got predominantly an agrarian economy, since the beginning of the planning process, its growth has occupied the centre stage. However, in recent years, the country has experienced a visible structural shift from agriculture to tertiary activities even bypassing the industrial sector. As a result, employment creation has perhaps become the biggest casualty.

In the wake of rapidly increasing population in India especially in rural areas, the pattern of growth should be such that it ensures adequate gainful employment opportunities. However, what is distressing to note is that in the last two decades, while the growth of gross domestic product (GDP) at the aggregate has been decent by the standards of developing countries, the country has witnessed sharp deceleration in the employment growth (Bhattacharya and Sakthivel, 2004; Sundaram, 2001a and 2001b). There are studies to reveal that growth has failed to improve employment conditions in the country, even though the growth rate of labour force has been quite low (Ghose 1999, Unni and Raveendran 2007).

From the point of view of the whole economy, employment growth is found to have slowed down from 2 per cent per annum during 1983-84/1993-94 (henceforth to be read 1983-93) to less than 1 per cent per annum during 1993-94/1999-2000 (henceforth 1993-99) (Planning Commission of India, 2001). In this respect, the performance of the primary sector is miserable, while it is moderate for manufacturing sector and somewhat satisfactory for service sector in the liberalized period of the Indian economy (Veni, 2004). Needless to say, there is a growing 'tertiarization' of the structure of production and employment in India ever since the inception of the planning process (Joshi, 2004). The tertiary sector may tend to play a catalytic role in employment generation and poverty alleviation in medium run. However, in the long run, simultaneous growth of all the sectors is aptly desirable (Joshi, 2004; Goldar, 2000; Ghose, 2006).

Given this backdrop, the present study attempts to develop a set of stylized facts about the attributes of agricultural income, productivity and employment and their linkages in the development process. Accordingly, the paper is organized into three sections. Section-II outlines the concept of employment, sources of data and the methodology as applied in the study. In section III, major trends in agricultural employment are analyzed and a comparative account of employment and income in agriculture at aggregate and disaggregate levels is presented. Labour productivity and its relationship with agricultural employment are presented in Section IV. Section V concludes the findings.

CONCEPTS, DATABASE AND METHODOLOGY

Employment can be defined from various perspectives depending upon the status of economic activities undertaken by the people. According to the National Sample Survey (NSS), the economic activity status is measured in terms of 'Usual Status Approach' or 'Current Status Approach'. While the former takes into consideration the number of persons in the workforce, the latter denotes the number of man-days. A person is treated as employed under the 'Usual Status Approach', if he/she had pursued gainful economic activity for a relatively longer period of time in the period immediately preceding one year (365 days) prior to the date of NSS survey. This is known as 'Usual Principal Activity Status'. On the other hand, if a person had spent relatively shorter time span in the period immediately preceding one year (365 days) prior to the date of survey, he/she is accounted under the head 'Usual Subsidiary Activity Status'. 'Principal Activity Status' and 'Subsidiary Activity Status' together constitutes 'Usual Activity Status Approach'.

The 'Current Status Approach', however, assigns a unique activity status for a person engaged in gainful economic activity for a period preceding one week or the previous day of the survey. If a person has pursued any one or more gainful economic activities for at least one hour during the preceding week, he/she is considered to be employed under 'Current Weekly Status' (CWS) approach. If a person has spent four hours or more during the previous day of the survey, he/she is considered to be employed under 'Current Daily Status' (CDS) approach.

The study is based on secondary data. The major sources of data are National Sample Survey Organization (NSSO) and Central Statistical Organisation (CSO). The study considers the post-reforms period from 1993-94 to 2004-05. During this period, NSSO's quinquennial survey results based on a large sample is available for employment. The entire period is divided into two periods–1993-1999 and 1999-2000/2004-05 (henceforth 1999-00). Wherever necessary, data related to the '80s are introduced for analysis. In the study, employment is defined on the basis of 'Usual Activity Status Approach'. The study estimates employment elasticity in agriculture as a ratio of percentage change in agricultural employment to the percentage in agricultural income. The relationship between employment growth and its determinants is presented following a pooled data model.

THE SCENARIO OF AGRICULTURAL INCOME AND EMPLOYMENT IN INDIA

Examining the trend of agricultural income (Table 1), it is observed that the contribution of agriculture to the country's GDP has declined dramatically in course of the study period. India has experienced a deceleration in agricultural income as a percentage of GDP from a high of about 30 per cent in 1993-94 to a low of about 20 per cent in 2004-05. The states have followed, by and large, the same trend. This decline is the highest in Kerala (from 25 per cent in 1993-94 to 11 per cent in 2004-05) followed by Orissa (from 34 per cent in 1993-94 to 21 per cent in 2004-05). The share of agricultural employment to total employment has also declined during the same period but not at the same rate as the fall in contribution of agricultural income to GDP. Employment in agriculture has decreased from about 60 per cent in 1993-94 to about 55 per cent in 2004-05. By and large, same trend is observed across all the major states excepting Bihar and Assam. Bihar has even experienced an increase in agricultural employment in 2004-05 (66 per cent) as compared to that in 1993-94 (65 per cent), while Assam has maintained almost the same level as it had in 1993-94. The overdependence on agriculture in Bihar could be attributed to lack of alternative sources of livelihood in the state. The other states which are largely dependent on agriculture are Madhya Pradesh followed by Karnataka, Maharashtra, Uttar Pradesh, Rajasthan and Gujarat. Kerala is least dependent on agriculture, where little over a third of the total employed are engaged in agriculture.

Table 1
Trends of Agricultural Income and Employment in Major States of India

	Share of Agricultural Income to GDP			*Share of Agricultural Employ-ment to Total Employment*		
States	*1993-94*	*1999-00*	*2004-05*	*1993-94*	*1999-00*	*2004-05*
Andhra Pradesh	30.11	24.06	21.10	62.07	59.13	54.41
Assam	34.79	31.13	25.63	57.36	52.25	57.01
Bihar	45.27	37.22	36.41	64.87	62.42	66.36
Gujarat	19.91	14.62	15.49	60.47	61.08	58.40
Haryana	41.81	33.25	27.66	59.64	58.14	54.49
Karnataka	32.95	26.92	17.00	63.51	62.29	60.42
Kerala	24.79	20.05	11.05	49.01	39.21	36.15
Madhya Pradesh	37.09	30.46	27.00	69.79	67.79	63.43
Maharashtra	17.88	14.70	10.14	62.10	61.53	59.56
Orissa	34.16	24.80	21.43	63.33	59.73	54.94
Punjab	45.80	39.74	36.31	61.40	58.00	53.44
Rajasthan	32.07	26.52	24.77	64.53	62.40	58.55
Tamil Nadu	22.37	16.46	12.13	54.38	51.61	49.36
Uttar Pradesh	37.94	34.27	30.48	64.58	59.46	58.54
West Bengal	27.75	23.36	18.65	45.21	43.00	43.82
All India	29.66	24.14	19.90	60.15	57.20	55.26

Source: Estimated from the data obtained from NSSO and CSO.

A further probe into the growth rate of income and employment provides interesting results (Table 2). At the country level, while the annual compound growth rate of income from agriculture declined from 2.82 per cent for the period 1983-93 to 1.27 per cent in 1999-04, the same for employment came down from 1.44 per cent to a low of 0.21 per cent in the respective periods. The situation in '90s was still worse in respect of employment growth. During this period, while income growth was 2.52 per cent, employment growth rate was mere 0.11 per cent. There was a situation of almost 'jobless growth' witnessed in Indian agriculture during this period. This is evident from the employment elasticity which is a measure of the employment intensity of growth of output in agricultural sector. The employment elasticity was mere 0.04 during the period 1993-99 as compared to 0.51 in the preceding decade. Some improvement was, however, noticed in this front in the next period (0.17 during 1999-04). Still it was indeed very low in the wake of increasing rural workforce.

Table 2
Annual Compound Growth Rates (ACGR) of Income and Employment, and Employment Elasticity in Agriculture in India

ACGR in Income			*ACGR in Agriculture Employment*			*Employment Elasticity*		
1983-93	*1993-99*	*1999-04*	*1983-93*	*1993-99*	*1999-04*	*1983-93*	*1993-99*	*1999-04*
2.82	2.52	1.27	1.44	0.11	0.21	0.51	0.04	0.17

Source: Estimated from the data obtained from NSSO and CSO.

This fall in employment in agriculture may have various ramifications. One possible explanation in support of this phenomenon could be that structural shift may have been underway in connivance with the growth of an economy. It is likely that as the economy grows, the share of non-agricultural sector to the country's GDP and its employment should rise (Hazra, 1991). However, in the context of the Indian economy, such a shift has not been evident particularly in the post-reform period, especially in the '90s. As studies indicate, the process of diversification of employment, which was in force from mid '70s to the late '80s, away from primary activities towards a variety of non-agricultural avenues, has tended to come under pressure and the growth rate of the latter has also slowed down considerably in '90s. This diversification, prior to '90s, was primarily attributed to significant acceleration in public expenditure in rural areas (Sen, 1996; Chandrasekhar and Ghosh, 2002). However, the '90s witnessed a policy shift for the worse in this regard (Jha, 2003).

In the wake of the shrinking job opportunities outside agriculture, the fall in employment in agriculture becomes a matter of serious concern. The labour absorptive capacity of agriculture in India is under stress for several reasons. The declining landman ratio, increasing marginalization of holdings, labour-saving cropping pattern adjustments, increasing mechanization of field crop operations, and increasing threat to domestic agriculture through open-door trade policy may be identified as some of the probable factors (Chadha and Sahu, 2002). As Jha (2006) asserts, agricultural product mix has changed significantly in the '90s and the impact of trade reforms is clearly evident in this. This phenomenon, for instance, has encouraged production of horticultural crops, which are primarily labour-saving in nature. Further, liberal import policy for some farm inputs like pesticides in the '90s has increased its use. There are evidences that these chemicals have replaced labour in certain regions of the country. Jha (2006) further argues that the price stickiness of agricultural produce, high agricultural wage and stagnating agricultural productivity are some other factors that have curtailed employment of labour in agriculture in some parts of the country.

At a disaggregate level, there are, however, mixed trends (Table 3). During '90s, in four out of fifteen major states, the annual compound growth rate of employment was negative. Those states were Andhra Pradesh, Kerala, Rajasthan and Tamil Nadu. In the later period (1999-04), Kerala continued to experience negative growth rate of employment. Orissa became another state which also experienced negative growth rate of employment in the later period. Interestingly, Punjab, one of the states known for being the frontrunner of 'Green Revolution' in the country, experienced a fall in annual compound growth rate of employment in recent years (1 per cent during 1999-04 as compared to 1.24 per cent in the preceding period). The other states which have experienced dismal performance are Tamil Nadu and Andhra Pradesh.

Given the scenario with regard to employment, agricultural income growth rates may provide some possible linkages. The study presents no uniform pattern of income changes across all the states. As many as nine states viz. Kerala, Karnataka, Maharashtra, Tamil Nadu, Madhya Pradesh, West Bengal, Rajasthan, Uttar Pradesh and Punjab registered a decline in the average compound growth rate of agricultural income during 1999-04. Amongst them, the former five states experienced even negative growth rate. On the other hand, from the rest, Gujarat experienced a significant increase in agricultural income in the recent years (6.59 per cent in 1999-04) followed by Bihar (3.76 per cent), Andhra Pradesh (3.12 per cent), Haryana (2.48 per cent), Orissa (2.34 per cent) and Assam (1.01 per cent). It is interesting to note that, in Orissa, while the income growth rate has turned out to be positive in the recent period from a negative growth rate in the preceding period, the employment growth has not only been abysmally low but also it has just been reversed. This employment-income relationship may be better understood from the estimates of employment elasticities for various states (Table 3).

Table 3
Annual Compound Growth Rates of Agricultural Income and Employment, and Employment Elasticity Across Major States of India

	ACGR in Agricultural Income		*ACGR in Agricultural Employment*		*Agricultural Employment Elasticity*	
States	*1993-99*	*1999-04*	*1993-99*	*1999-04*	*1993-99-*	*1999-04*
Andhra Pradesh	1.59	3.12	-1.06	0.87	-0.66	0.28
Assam	0.22	1.01	1.93	2.77	8.85	2.75
Bihar	0.69	3.76	1.06	3.4	1.54	0.9
Gujarat	2.43	6.59	2.3	2.17	0.95	0.33
Haryana	1.94	2.48	0.93	3.24	0.48	1.3
Karnataka	4.06	-2.72	1.3	1.55	0.32	-0.57
Kerala	1.97	-4.32	-2.66	-0.42	-1.35	0.1
Madhya Pradesh	2.82	-0.35	1.14	1.22	0.41	-3.46
Maharashtra	2.83	-2.09	1.65	2	0.58	-0.96
Orissa	-1.13	2.34	0.51	-0.37	-0.45	-0.16
Punjab	2.32	1.73	1.24	1	0.54	0.58
Rajasthan	4.83	2.8	-0.65	3.41	-0.13	1.22
Tamil Nadu	1.32	-1.75	-0.03	0.39	-0.02	-0.22
Uttar Pradesh	2.92	1.16	0.43	3.01	0.15	2.58
West Bengal	4.08	1.91	0.78	2.34	0.19	1.22
All India	2.52	1.27	0.11	0.21	0.04	0.17

Source: Estimated from the data obtained from various Reports of NSSO.

It is disturbing to note that not only the employment elasticity is very low in most of the states, but also five of them have registered negative employment elasticity in both the periods under study. While in '90s, Andhra Pradesh, Kerala, Orissa, Rajasthan and Tamil Nadu experienced negative elasticity, in the later period, Karnataka, Madhya Pradesh, Maharashtra, Orissa, Tamil Nadu experienced negative elasticity. The situation in the recent period indicates that only five states had elasticity greater than one. Those states were Assam, followed by Uttar Pradesh, Haryana, West Bengal and Rajasthan.

Negative employment elasticity may need further explanation. As it is evident, negative employment elasticity in any state is reached at either due to negative income growth or due to negative employment growth of that state. During the period 1993-99, barring Orissa, in four other states, where negative employment elasticity was registered, it was due to negative employment growth. However, in the later period, barring Orissa, the other four states had negative income growth leading to attainment of negative employment elasticity. In these four states, employment growth has increased during the reference period. This may have serious implications in terms of labour productivity, wages and welfare of the agricultural workers (Jha, 2006). The implication of this relationship with respect to labour productivity is presented in the following section.

LABOUR PRODUCTIVITY AND EMPLOYMENT IN AGRICULTURE

Labour productivity is one of the various dimensions of productivity that plays an important role in determining the level of aggregate output. Needless to mention that improvement in labour productivity is warranted without, however, a compromise with employment in agriculture especially in a labour surplus rural economy like that of India. Though labour productivity can be defined in various ways, in the present study, it is defined as agricultural GDP at 1993-94 price divided by the number of persons employed in agriculture on UPSS basis. The spatial and temporal trend in labour productivity is, accordingly, presented in Table 4.

The results show wide diversity in labour productivity across states in the periods under study. In all the three periods, Punjab followed by Haryana topped the list. At the other end, in 1993-94, Bihar registered the lowest labour productivity followed by Orissa, while in mid-period, the order was just reversed. However, in 2004-05, Maharashtra became the low productive state followed by Bihar and Orissa. It is interesting to note that labour productivity of agriculture in Punjab and Haryana is more than four times than that in Bihar in the entire study period. Between the periods 1993-94 and 1999-00, decline in labour productivity was registered in three states namely Orissa, Bihar and Assam. That is the reason why the annual compound growth rate of labour productivity during 1993-99 was recorded negative for these three states. In the next period, decline in labour productivity was registered in as many as 10 states leading to attainment of negative growth rates in all of them. The states which experienced acceleration in labour productivity were Punjab, Andhra Pradesh, Gujarat, Bihar and Orissa.

Table 4
Trends of Average Labour Productivity in Agriculture Across Major States of India

	Average Labour productivity (Rs)			*ACGR in Labour Productivity*	
States	*1993-94*	*1999-00*	*2004-05*	*1993/94 -1999/00*	*1999/00-2004/05*
Andhra Pradesh	8474.74	9932.94	11441.24	2.68	2.87
Assam	11539.65	10423.58	9655.20	-1.68	-1.52
Bihar	7516.55	7352.54	7762.42	-0.37	1.09
Gujarat	8939.82	9006.74	11865.01	0.12	5.67
Haryana	24519.20	26024.89	25709.34	1.00	-0.24
Karnataka	10220.69	12010.73	9427.43	2.73	-4.73
Kerala	11788.99	15581.80	12213.92	4.76	-4.75
Madhya Pradesh	9643.20	10640.33	9806.45	1.65	-1.62
Maharashtra	8970.43	9616.01	7672.55	1.17	-4.42
Orissa	7726.79	7002.48	8194.01	-1.63	3.19
Punjab	29068.11	30974.39	32675.17	1.06	1.07
Rajasthan	8363.46	11540.59	11516.67	5.51	-0.04
Tamil Nadu	8715.67	9443.91	8333.33	1.35	-2.47
Uttar Pradesh	9884.86	11449.98	10582.23	2.48	-1.56
West Bengal	12405.83	15053.41	15025.61	3.28	-0.04
All India	177777.97	196054.31	191880.58	1.64	-0.43

Source: Estimated from the data obtained from NSSO and CSO.

The relationship between labour productivity in agriculture and employment in agriculture, in the context of India, can be formally presented with the help of the following regression results (Table 5). Taking the entire period into account, the study finds an inverse relationship between labour productivity growth in agriculture and agricultural employment growth. This corroborates with the notion that higher labour productivity leads to lower scope for employment (Landmann, 2004). As expected, higher agricultural income growth leads to higher agricultural employment growth. An attempt was also made to see if there is any change in the impact of these two variables on employment by splitting the entire period into two (1994-95/1999-2000 and 1999-2000/2004-05). However, no significant difference in the result was observed. Thus, that is not reported in the paper.

Table 5
Regression Results between Employment in Agriculture and its Determinants: OLS (Pooled Model) Estimates for the Period 1994/95-2004/05

	Model	
Variables	*Un-standardized Coefficients*	*Standardized Coefficients*
Constant	0.00037	
	(0.09376)	
AGRAI	0.88103*	1.558
	(105.7722)	
AGALP	-0.53513*	-1.816
	(-123.306)	
R^2		0.989473
Adjusted R^2		0.989343
F-value		(2, 162) 7574.665*
N		165

Note: AGRAI: Annual Growth Rate of Agricultural Income; AGALP: Annual Growth Rate of Agricultural Labour Productivity; * significant at 1 per cent level; Figures within parentheses represent values of the t-statistic.

In India, a comparison between labour productivity and employment can be attempted by clustering the states under various alternative situations viz. positive labour productivity growth versus positive employment growth, negative labour productivity growth versus positive employment growth, positive labour productivity growth versus negative employment growth, and negative labour productivity growth and negative employment growth (Box 1). While during the period 1993-99, eight states showed positive employment growth along with a positive labour productivity growth, in the later period, only four states had to experience that situation. The states falling under this category during 1993-99 were Gujarat, Haryana, Karnataka, Madhya Pradesh, Maharashtra, Punjab, Uttar Pradesh and West Bengal. Andhra Pradesh, Bihar, Gujarat and Punjab were the four states which experienced the same situation in the next period. As against this, during 1999-04, as many as nine states namely Assam, Haryana, Karnataka, Madhya Pradesh, Maharashtra, Rajasthan, Tamil Nadu, Uttar Pradesh and West Bengal experienced a situation of negative labour productivity growth leading to positive employment growth versus only three states in the preceding period (Assam, Bihar and Orissa). Another disturbing feature was positive labour productivity growth versus negative employment growth. This was experienced by Andhra Pradesh, Kerala, Rajasthan and Tamil Nadu during the first period, whereas Orissa remained the only state to face this situation in the later period. Kerala was the only state that fell under the last category i.e. negative labour productivity growth and negative employment growth.

Table 6
Growth (EMP) during 1993-99 and 1999-04

+ve LPR and +ve EMP:		
1993-1999	:	Gujarat, Haryana, Karnataka, Madhya Pradesh, Maharashtra, Punjab, Uttar Pradesh, West Bengal
1999-04	:	Andhra Pradesh, Bihar, Gujarat, Punjab
-ve LPR and +ve EMP :		
1993-99	:	Assam, Bihar, Orissa
1999-04	:	Assam, Haryana, Karnataka, Madhya Pradesh, Maharashtra, Rajasthan, Tamil Nadu, Uttar Pradesh, West Bengal
+ve LPR and –ve EMP :		
1993-99	:	Andhra Pradesh, Kerala, Rajasthan, Tamil Nadu
1999-04	:	Orissa
-ve LPR and –ve EMP:		
1993-99	:	nil
1999-04	:	Kerala

This relationship between labour productivity and employment does not present a uniform pattern. The notion of a tradeoff between employment and productivity growth (Landmann, 2004) finds its application in 10 states in recent period. In '90s, the situation was different as eight states showed positive employment growth when the growth of average productivity of labour was reported to be positive. In India's labour surplus rural economy, an increase in labour productivity without compromising with employment may seem to be a better bet. As Bhaduri (2005) points out, especially in poor countries with technological backwardness and other handicaps regarding international marketing, productivity increase along with higher employment is a necessity, more for augmenting the availability of total output than international competitiveness.

However, that situation is not apparently observed in Indian agriculture in recent years. Only four states of the country viz. Andhra Pradesh, Bihar, Gujarat and Punjab have experienced such a situation during 1999-04. Kerala is the other state that experiences negative employment growth accompanied by negative labour productivity growth. On the contrary, in nine other states, higher employment has been registered only with a fall in labour productivity. Orissa's situation appears to be still miserable as that has experienced decline in employment with an increase in labour productivity.

CONCLUSION

It is evident from the study that in India, employment in agriculture is on the decline in the post-reform period along with the deceleration in agricultural income. The situation in '90s was worse in respect of employment growth. There was a situation of almost 'jobless growth' in Indian agriculture during this period as evident from the employment elasticity.

By and large, the same trend is witnessed across major Indian states. The states which are largely dependent on agriculture are Bihar, Madhya Pradesh, Karnataka, Maharashtra, Uttar Pradesh, Rajasthan and Gujarat. Kerala, on the contrary, is the least dependent state. Not only the employment elasticity is very low in most of the states, but also five of them have registered negative employment elasticity in the entire period. While in '90s, Andhra Pradesh, Kerala, Orissa, Rajasthan and Tamil Nadu experienced negative elasticity, in the later period, Karnataka, Madhya Pradesh, Maharashtra, Orissa, Tamil Nadu experienced negative elasticity.

A further probe into the relationship between employment growth and productivity growth indicates that while in '90s, Gujarat, Haryana, Karnataka, Madhya Pradesh, Maharashtra, Punjab, Uttar Pradesh and West Bengal experienced positive employment growth along with a positive labour productivity growth, in the later period, only four states namely Andhra Pradesh, Bihar, Gujarat and Punjab had to experience that situation. As against this, Assam, Haryana, Karnataka, Madhya Pradesh, Maharashtra, Rajasthan, Tamil Nadu, Uttar Pradesh and West Bengal experienced a situation of negative labour productivity growth leading to positive employment growth in 1999-04. Assam, Bihar and Orissa had experienced this situation in the preceding period. Andhra Pradesh, Kerala, Rajasthan and Tamil Nadu experienced positive labour productivity growth with negative employment growth during 1993-99. Orissa is the only state to face the same in the recent period. Kerala, on the other hand, is an exception, which has experienced negative employment growth coupled with negative productivity growth.

While it is perceived that a fall in employment in primary sector, including agriculture, is a natural outcome of economic progress with the possibility of a structural shift, this is not evident in Indian economy. In the wake of shrinking job opportunities outside agriculture, fall in employment elasticity in agriculture is a matter of serious concern. Besides, in India's labour surplus rural economy, an increase in labour productivity without deceleration in employment may be a necessity, which is seriously missing in Indian agriculture. Unless adequate employment is created in the non-agricultural sector, a decline in agricultural employment may prove to be self-defeating.

REFERENCES

Bhaduri, A. (2005), "Macroeconomic Policies for Higher Employment in the Era of Globalization" Employment Strategy Papers, Employment Analysis Unit, Employment Strategy Department, ILO, Series No. 2005/11.

Bhattacharya, B.B. and Sakthivel, S. (2004), Economic Reforms and Jobless Growth in India in the 1990s. Working Paper Series No. E/245/2004, Institute of Economic Growth, New Delhi.

Chandrasekhar, C. P. & J. Ghosh (2002), "The Market that Failed: A Decade of New-Liberal Economic Reforms in India", Left World, New Delhi.

Chadha, G. K. and Sahu, P. P. (2002), Post-reform Setbacks in Rural Employment: Issues that Need Further Scrutiny, *Economic and Political Weekly*, 37 (21).

Ghose A. K. (1999), Current Issues of Employment Policy in India, *Economic and Political Weekly*, 34 (36).

Ghose A. K. (2006), Economic Growth and Employment in Labour-Surplus Economies, *Economic and Political Weekly*, 41 (31).

Goldar, B. (2000), Employment Growth in Organised Manufacturing in India, *Economic and Political Weekly*, 35 (14).

Government of India (2001), " Planning Commission Annual Report 2001", New Delhi.

Hazra, S. (1991), "Employment in India's Organised Sector", *Social Scientist*, 19 (7).

Hirschman, A. O. (1958), "Strategy of Economic Development", Yale University Press, New Haven, New York.

Jha, B. (2006), "Employment, Wages and Productivity in Indian Agriculture", IEGWorking Paper Series No. E/266/2006.

Jha, P. (2003), "Issues Relating to Employment in India in the Era of Globalisation", *Social Scientist*, 31(11/12).

Joshi S. (2004), Tertiary Sector-Driven Growth in India, *Economic and Political Weekly*, 39 (37).

Jorgenson, D.G. (1961), "The Development of a Dual Economy", *Economic Journal*, 71.

Lewis, W.A. (1954), "Economic Development with Unlimited Supplies of Labour", *The Manchester School of Economics and Social Studies*, 22.

Mellor, J. W. (1976), "The New Economics of Growth: A Strategy for India and the Developing World", Ithaca, New York: Cornell University Press.

National Sample Survey Organisation (1995), "Employment and Unemployment in India, 1993-94, NSS Fiftieth Round, NSS Report", No. 409, NSSO, New Delhi.

National Sample Survey Organisation (2001), "Employment and Unemployment in India, 1999-00, NSS Fiftieth Round, NSS Report", No. 458, NSSO, New Delhi.

National Sample Survey Organisation (2006), "Employment and Unemployment in India, 2004-05, NSS Fiftieth Round, NSS Report", No. 516, NSSO, New Delhi.

Landmann, O. (2004), "Employment, Productivity and Output Growth", Employment Strategy Papers Employment Analysis Unit, Employment Strategy Department, ILO, Series No. 2004/ 17.

Scitovsky, T. (1954), "Two Concepts of External Economies", *Journal of Political Economy*, 62.

Rosenstein-Rodan, P. N. *(1943), "Problems of Industrialization of Eastern and South-eastern Europe", Economics Journal*, 53.

Sen, A. (1996), "Economic Reforms, Employment and Poverty: Trends and Options", *Economic and Political Weekly*, 31 (35).

Sundaram, K. (2001a), Employment-unemployment Situation in the Nineties-Some Results from NSS 55th Round Survey, *Economic and Political Weekly*, 36 (11).

Sundaram, K. (2001b), Employment and Poverty in 1990s–Further Results from NSS 55th Round Employment-unemployment Survey, 1999-2000, *Economic and Political Weekly*, 36 (32).

Unni J., Raveendran G. (2007), Growth of Employment (1993-94 to 2004-05): Illusion of Inclusiveness?, *Economic and Political Weekly*, 42 (03).

Veni L. Krishna (2004), Changing Employment Scenario in India–Some Reflections since Liberalization, *The ICFAI Journal of Managerial Economics*, 2 (02).

wise and region-wise differences is of course corroborated by a number of studies (Krishnaji 1975). The impact of green revolution on the yield of certain crops (wheat, and to a lesser extent maize etc.) in the initial period of green revolution was noticeable increasing the regional disparities both in the yield and nature of crops. Moreover, agrarian strategy, at least during this period, also aimed at maximizing yield per unit land in developed areas through a connected application of all inputs. Paradoxically, however regional disparities in yield of all crops taken together, measured by the coefficient of variation, did not increase significantly at the district, region or state level during the 1960s or still later periods (Bhalla and Alagh, 1979; Dev, 1985; Sawant, 1987). These results appear to hold because of three main reasons. Firstly, several regions (like Gujarat, Rajasthan, Haryana etc.) which would have otherwise experienced still lower growth rate in productivity, experienced fairly rapid growth rates in the area under high valued cereals (Krishnaji,1975; Sawant,1987; Reserve Bank of India, 1984). Secondly, compensating price movements in the case of slower growing non-food growing crops also tended to mitigate value differences in yield per acre (Misra, 1985; Sawant, 1987). The net outcome of these factors was such that between the early 1960s and the late 1970s, the relative ranking of different states in the value of crop production per hectare remained more or less invariant, with West Bengal and Tamil Nadu occupying the first two places (Misra, 1985). Thirdly, several regions with initial low and medium productivity levels experienced rapid growth rates in overall yields, whereas many of the regions the fertile alluvial or coastal or delta plains experienced low growth rates in land productivity[1]. On the other hand, several regions, with high initial levels of land productivity, showed low growth rates in productivity up to the late 1970s (Mahendra Dev, 1985)[2]. During 1962-65 and 1970-73, districts experiencing high growth rates for all crops were concentrated in Punjab, Haryana, Uttar Pradesh, and Rajasthan with these states accounting for 36 of the 50 high growth districts. Gujarat, Karnataka and Tamil Nadu accounted for another 10 of the high growth districts. Only one out of 60 districts was located in the central and eastern regions. At the other end, 133 out of the 289 districts studied experienced growth rates below 1.5 per cent per annum. Most of these were located in Andhra Pradesh, Bihar, Karnataka, Gujarat, Madhya Pradesh, Orissa and Uttar Pradesh (Bhalla and Alagh, 1979). Between 1969-72 and 1981-84, state level results sustained high growth rates (over 3%) in Punjab and Haryana and an acceleration to high growth rates in a few states in the North, West and South (e.g., Maharastra, Gujarat, Uttar Pradesh and Andhra Pradesh). Growth rates in Karnataka and Rajasthan show some deceleration compared to the 1960s, but remained above average. For a similar period (1970-73 to 1980-83), Bhalla and Tyagi (1989) show that while in the decade ending in the triennial 1968-69 to 1970 only 71 districts (out of 284), accounting for 25% of gross cropped area, exprienced agricultural output growth rates above 3.5 %, by the end of the decade, the number had increased to 95 districts, covering nearly 40% of the area. It has been noted among the significant features of the growth pattern that emerged during this period was the continuation of very high growth rates in most of the original green revolution areas, the appearance of a large new set of the high growth districts in the major states (e.g., Uttar Pradesh, Andhra Pradesh, Gujarat and Maharastra) together with the unambiguous worsening of conditions in Bihar, Tamil Nadu and West Bengal and the down ward shift of some districts in Rajasthan and Madhya Pradesh (Bhalla, 1988, pp. 4-5). For the decade of the 1960s (1960-63 to 1970-73), except for a few regions in the North and North-West, South, South-East and Bengal, all other regions (a total of 27 regions out of 49 regions) experienced negative growth in labour productivity (Kundu and Raza, 1982). This picture is also confirmed by a more disaggregated analysis by Bhalla and Alagh. The level of agriculture yield is known to be significantly correlated with the key demographic variable. But the strength of this relationship declines with time. The level of yield is also related to the level of rainfall and negatively related to the proportion of rainfall during June-September (Bardhan, 1983). Further, the environmental demographic variables are also related to a number of structural variables. Further, between 1962-65 and

1975-78, a number of regions with low initial levels of labour productivity, low or medium rates in this variables[3], or otherwise experienced declining labour productivities (Mahendra Dev, 1986)[4]. Thus, several regions, chiefly in the eastern, central and western parts appear to be caught in the low productivity low growth tap.The nature of 'associated' changes such as changes in land, lease and labour markets—symptomatic of an uneven penetration of capitalism in agriculture, is the subject matter of a number of detailed investigations, and is not pursued here.(See, for instance, Bharadwaj and Das, 1975; Bhardhan and Rudra, 1978; Byers, 1981; Khasnabis and Chakarvarty,1982,; Harris,1982; Chatopadhyay and Bhattacharya, 1984). But Bhalla's conclusion, mirrors Byers's analysis above on a more macro plane.

In the recent period, a number of studies (for instace, Ahluwalia 2000 and 2002; Nagraj, Varoudakis and Veganzous 1998; Rao, Shand and Kalirajan 1999; Shand and Bhide 2000) have observed that the regional disparities in India has widened, especially during the 1990s. However, these studies have used pre-revised state domestic product (SDP) data. Further their analyses do not cover adequately the post-reform period, especially the late 1990s when the growth rate slackened at the aggregate as well as at the regional levels. Bhattacharya and Sakthivel (2004), carried a State-Wise study on the basis of State Domestic Product and found that the growth performance and structural changes in the domestic product of Indian states in the last two decades has been uneven across states. While advanced industrial states have tended to leapfrog in the reform years, other states have lagged behind. The regional disparities in growth rates become sharper in terms of per capita income. C.H. Hanumantha Rao (2006), conducted a inter-state study in the post reform period. He found the disparities continuously going to increase by the large variabilities in the contribution made by the secondary and tertiary sectors.

Thus, in the light of the studies reviewed, the present study would concentrate on examining the trends of agriculture development within districts in Haryana with the help of various agricultural variables. Hence, this empirical work would expect contribute to developing a better understanding of the different variables regarding the agricultural development. The main objective of the present study is, therefore, to analyse the district–wise pattern of agricultural development in Haryana since 1992 and onward focusing on inter-district variations in level and pattern of agricultural development. In the present study the following indicators have been taken to analyse the level and pattern of agricultural development in Haryana. (1) Agricultural resource base (rainfall/soil fertility, per capita net area sown, percentage irrigated area, double cropped area, area under cash crops). (2) Adoption of new technology (HYV, fertilizer, plant protection, mechanization and area counter doubled). (3) Agriculture infrastructure facilities (credit, communication, marketing, storage capacity, rural electrification and soil setting).

SPATIAL PATTERN OF AGRICULTURAL DEVELOPMENT IN HARYANA

Agriculture is the key to restructuring the economy and hence the role of agriculture sector in economic development of different districts seems to be the most crucial. During the eighties and the nineties, there had been a significant improvement in the technology in agriculture sector with spread of high yielding varieties and wider expansion of irrigation. The barriers of land tenure system, use of primitive technology and lack of infrastructure for raising productivity have been overcome to a considerable extent. In view of these, it would be interesting to analyse the extent to which these structural transformations have brought about changes in inter-district patterns of agricultural development. For this purpose, we have selected five indicators to represent agricultural development of different districts. These indicators are agricultural production per hectare of net area shown (A_1), consumption of power per hectare in agriculture (A_2), percentage of net irrigated area to net area sown (A_3), consumption of fertilizer per hectore of gross cropped area (A_4)

and cropping intensity (A_5) shows a combined influence of agricultural inputs. The correlation matrix of these indictors of agricultural development for the years 1992-93 and 1904-05 are given in Tables 1 and 2. As is evident from the above tables, all of the five indicators of agricultural developed are positively associated with each other in both the years (1992-93 and 2004-05). In respect of 1992-93, the first principal component explains 77.014 per cent of the total variation. The equation for the composite index of agricultural development (Z_1) stands as:

$$Z_1 = 0.9346\ A_1 + 0.6534\ A_2 + 0.8781\ A_3 + 0.7922\ A_4 + 0.9862\ A_5$$

Table 1
Inter Correlation Matrix of Agricultural Development Indicators 1992-93

	A_1	A_2	A_3	A_4	A_5
A_1	1.00				
A_2	0.74	1.00			
A_3	0.93	0.73	1.00		
A_4	0.96	0.72	0.95	1.00	
A_5	0.72	0.87	0.66	0.71	1.00

Number of observation = 16.

Table 2
Inter Correlation Matrix of Agricultural Development Indicators 2004-05

	A_1	A_2	A_3	A_4	A_5
A_1	1.00				
A_2	0.87	1.00			
A_3	0.86	0.95	1.00		
A_4	0.87	0.97	0.96	1.00	
A_5	0.74	0.57	0.61	0.44	1.00

Number of observation = 16.

where, A_1, A_2, A_3, A_4 and A_5 are. Indicators in the standardized form.

The coefficients of correlation of Z_1 with each of the selected variables are positive and their estimate values are 0.93, 0.65, 0.88, 0.79 and 0.98 respectively. Moreover, in respect of 2004-05, the principle component explains marginally over 82.92 per cent of the total variation. The equation of its composite index (Z_2) works out as:

$$Z_2 = 0.9059\ A_1 + 0.6713\ A_2 + 0.8959\ A_3 + 0.6313\ A_4 + 0.8825\ A_5$$

where, A_1, A_2, A_3, A_4 and A_5 denote values of the selected variables in the standardized form.

The coefficients of correlation of Z_2 with each of the selected variables are positive and their values stand as .90, 0.67, 0.89, 0.63 and 0.88, respectively. The estimates of composite indices of agricultural development for different districts separately for the years 1992-93 and 2004-05 are given in Appendix. The classification of districts according to composite index of agriculture development into high, medium high, medium low and low categories is depicted in Table 3.

Table 3
Classification of Districts according to Composite Index of Agricultural Development

Category of Development	*1992-93*	*1904-05*
High (1.1203 and above)	Kethal, Karnal, Kurukshetra (3)	Kethal, Karnal, Kurukshetra,Hissar (4)
Medium high (.5411 to < 1.1203)	Hissar,Panipat, Ambala, Jind, Yamuna Nagar (5)	Sirsa,Ambala,Jind, Yamuna Nagar Bhiwani (5)
Medium low (-.4506 to < .5411)	Bhiwani,Rohatak, Sirsa (3)	Gaurgawn, Panipat,, Sonipat, Fridabad (4)
Low (Below -.4506)	Gurgawn,Rewari, Mahendragarh, Fridabad, Sonipat (5)	Rewari,Mahendragarh, Rohatak (3)

Analysis attempted here is essentially two fold. First, inter-temporal analysis is sought through examining the upward and downward movements of different districts in level of agricultural development during 2004-05 over the year 1992-93. In 2004-05, the 3 districts are still found to be underdeveloped or developing in the matter of agricultural development. It is further observed that against the total number of 16 districts, there are 4 districts which showed marginal inter-category movements in levels of their agricultural development during 2004-05 over the year 1992-93. Some districts belonged to the category of medium low level of agricultural development during 1992-93 but during 2004-05 one more district attained higher level of agricultural development. Although absolute values of composite indices of agricultural development showed an improvement in most of the cases.

As a result of the aforesaid changes, there is an increase in the state level arithmetic mean of composite index of agricultural development for all the 16 districts from 0.4150 in 1992-93 to 0.4707 in 2004-05 as would be evident from the Appendix. Moreover, the number of districts having composite indices of agricultural development above the state average (mean) increased from 7 in 1992-93 to 13 in year 2004-05. Besides, coefficient of variance as a measure of inter-district disparities in levels of agricultural development shows increase from 434.0505 to 460.1652 during this period.

So far the inter-category comparisons of agricultural development is concerned, it is clear from Table 2. 3, that in 1992-93 the number of districts falling in the category of medium low and low was as high as 8, where, those belonging to high and medium high categories were 8 also. But there seems to be some definite improvement in agriculture situation during 1992-93 to 2004-05. This is supported by the fact that the number of districts falling in medium low and low categories considerably reduced to 7 and those constituting a high and medium high categories of its development simultaneously increased to 9. The improvement in agricultural situation which took place mainly in the districts of medium low level of its development during 2004-05 might be a outcome of better and effective implementation of various area specific programmes formulated for the development of backward areas during nineties.

Thus, three interesting findings noticeable from the forgoing analysis are (a) an upward movement of certain districts in level of agricultural development, (b) an increase in values of its composite indices in sizeable number of districts, and (c) reduction in inter-district disparities in levels of agricultural development. These findings, in turn, provide a substantial support to further conclude that there has been some definite improvement in inter-district pattern of agricultural development during 2004-05 as compared to 1992-93.

CONCLUSION

To conclude, the pattern of agriculture development among districts in the state largely goes in the favour of more developed districts, there seems considerable unevenness in the level of inter-district disparities of irrigation benefit in Haryana. The attempts at reduction of such lopsidedness do not appear to have been effective. Such disparities have covered divergences in relative agricultural prosperity among districts and have, thus, caused simmering discontentments in some cases. A shift in the cropping pattern in favour of commercial crops is indicated. Though this is a welcome shift, however, the yield rates in many districts have remained sticky over this period in some Kharif and Rabi crops is unattractive from the point of total output of agriculture. Hence, the development of agriculture in the state reveals the need for pulling up the agricultural economy and setting it on a path of sustained growth. In this connection, the identification and stimulation of real policy variables warrant priority in the agricultural development strategy of the state in the changing scenario of Globalization.

NOTES

1. The regions which experienced growth rates in yield exceeding 2.5 per cent between 1962-65 and 1975-78 are Gujarat (Dry Areas), Gujarat (Sourastra), Haryana (East and West), Punjab (North and South), Jammu and Kashmir (Mountainous and outer Hills), Karnataka (Inland Northern), Maharastra (eastern), Rajasthan (Western and Eastern), Uttar Pradesh (Western), Tamil Nadu (Costal Northern)–only the Tamilnadu (Costal Northern) region had an initial productivity level greater than Rs. 1300 per hectare (Mahendra Dev).
2. These reasons comprised virtually the whole of the Eastern part of the country, Andhra Padesh (Inland Northern and Southern), Gujarat (eastern and Southern) and Tamil Nadu (Costal Inland).
3. Jammu and Kashmir (Jhelum), Uttar Pradesh (Inland Northern), Bihar (Northern and Southern), Rajasthan(Southern), Maharastra (Inland Eastern), Madhya Pradesh (Inland Eastern).
4. Gujarat (Northern plains), West Bengal (Himalyain), West Bengal (Eastern Plain), Gujarat (Southrn plains), Madhya Pradesh (Eastern), Orissa (Southern), Madhya Pradesh).

REFERENCES

Books

Ahluwalia, Montek S. (2002), 'State Level Performance Under Economic Reforms in India' in Anne O Krueger (ed), Economic Policy Reforms and the Indian Economy, Oxford University Press, New Delhi, pp. 91-125.

Alagh, Y. K., Bhalla G. S. and Kashyap, S. P. (1980), 'Structural Analysis of Gujarat, Punjab and Haryana Economies, Delhi.

Atkinson, A.B., (1983), The Economics of Inequalities', Oxford University Press, New Delhi.

Bhalla, G. S. and Tyagi, D. S. (1989), Patterns in Indian Agricultral Development, A District Level Study, Institute for Studies in Industrial Development, New Delhi.

Bhalla, G. S. and Alagh, Y. K. (1979), Performance of Indian Agriculture: A District-wise Study, New Delhi.

Bhat, L S. *et al.*, eds. (1982), Regional Inequalities in India: An Inier-State and Intra-State Analysis, Society for the Study of Regional Disparities, New Delhi.

Dubey, K. H. (1988), 'The role of Decentralised Planning in Removing Inter-district Imbalances', Paper presented at National Seminar on. Decentralised Planning, State, Planning Institute, Lucknow.

Govt. of Haryana, Economics of Farming in Haryana (1983-84), Published by the Economic and Statistical Organization, Chandigarh.

Hinderink, J. and J. J. Sterkenburg (1978), 'Spatial Inequality in Under-developed Countries and the Role of Government Policy, Tjdschrift, Veer Economische in Social Graphapic, 69, 5-16.

Kerr, J., Pangae G., Lokur V. and George P. J. (1999), 'Effects of Watershed Project Subsidies on Soil and Water Conservation Investment in India's Semi Arid Tropics, In using Incentives in Soil Conservation, ed. D. Sandres, P. Huszar, S. Sombatpanit, and T. Enters. Enfield, N. H., U. S. A., Science Publishers.

Kalavalli, S. (1998), 'A Review of Approaches to Watershed Development in India: A Report submitted to the National Centre for Agriculture Economics and Policy Research, Indian Council of Agricultural Research, New Delhi.

Kundu, A. and M. Raza (1982), Indian Economy: The Regional Dimension, Centre for the Study of Regional Development, Jawaharlal Nehru University, New Delhi.

Nair, K.R.G. (1985), 'Inter-state Income Differentials in India 1970-71 to 1979-80', in Mishra G. P, ed. Regional Structure of Development and Growth in India, Vol. I and II, Ashish Publishing Hous, Delhi.

Somra, S. S. (1999), "Issues in Economic Development", RBSA Publication Jaipur.

Articles

Ahluwalia, Montek S. (2000), 'Economic Performance of States in Post-Reforms Period', *Economic and Political Weekly,* May, 6, pp. 1637-48.

Bardhan, (1983), 'Regional Variation in the Rural Economy', *Economic and Political Weekly,* XVIII, 30, July, pp. 1319-1334.

Barro, Robert J. (1991), Economic Growth in a Cross Section of Countries', *Quarterly Journal of Economics,* 106, 2, May, pp. 407-43.

Bhattacharya, B. B. and Sakthivel, S. (2004), 'Regional Growth and Disparities in India: Comparision of Pre- and Post-Reforms Decades', *Economic and Political Weekly,* March, 6, pp. 1071-1077.

Choudhary, M. K. and D. P. Gupta (1987), 'Regional Disparities in Income from Agriculture in Haryana, Kautiyla, *Journal of the Haryana Economic Association,* March.

Dadibhavi, R. V. and S. T. Bagalkoti (2006), 'Reforms and Regional Inequalities in India', *The Indian Economic Journal,* Vol. 54, Number 02, July-September, pp. 21-38.

Deaton, Angus (2003), Regional Poverty Estimates for India,1999-2000' Research Programme in Development Studies, Princeton University (mimeo) *www.prinston .edu/~rpds.*

Deaton, Angus and Jean Dreze (2002), 'Poverty and Inequality in India–A Re-Examination', *Economic and Political Weekly,* September 7, pp. 3729-48.

Haryana, Y. and V. W. Ratan, (1970), 'Agricultural Productivity Differences among Countries, *American Economic Review,* Vol. LX, No. 5, Dec.

Krishanaji, N. (1975), 'Inter Regional Disparities in Per Capita Production and Productivity of Foodgrains: A Preliminary Note on Trends', *Economic and Political Weekly,* Special Number, pp. 1377-1385.

Nash, M. (1963), 'Introduction Approaches to the Study of Economic Growth, *Journal of Social Issues,* 19,1-5.

Pal, M. N. (1975), Regional Disparities in the Level of Development in India', *Indian Journal of Regional Sciences,* 7, 35-51.

Rao, C.H. Hanumantha (2006), 'Growing Regional Disparities in Development in India-Post Reform Experience and Challenges Ahead', *The Indian Economic Journal,* Vol. 54, Number 01, April-June, pp. 52-61.

Rao, C.H. Hanumantha and Mahendra Dev (2003), 'Economic Reforms and Challenges Ahead—An Overview', *Economic and Political Weekly,* Specil Issue on Andhra Pradesh, March, 22-29, pp. 1130-41.

Rao, C.H. Hanmuntha (2000), 'Watershed Development in India: Recent Experience and Emerging Issues, *Economic and Political Weekly,* November 4, pp. 3943-3947.

Rao, M. Govind, R.T. Shad and K.P. Kalirajan (1999), 'Convergence of Income Across Indian States—A Divergent View', *Economic and Political Weekly,* March 27.

Shand, Ric and S Bhide (2002), 'Soure of Economic Growth—Regional Dimentions of Reforms', *Economic and Political Weekly,* October 14, pp. 3747-57.

Somra, S. S. (2003), "Inter-State Imbalances and Economic Reforms in India" in the *Journal of Sardar Patel Institute of Economic and Social Research,* Ahmedabad *'anvesak'* , July-Dec. 2002, Volume 32, No. 2, pp. 77-88

Sundram, K. and Suresh D. Tendulkar (2003), 'Poverty in India in the 1990s: Revised Result for All-India and 15 Major States for 1993-94', *EPW,* November 15-22.

Williamson, J.G. (1965), 'Regional Inequality and the Process of National Development: A Description of Pattern', *Economic Development and Cultural Change,* 4, Part 2, 3-84.

APPENDIX

Selected indicators of agricultural development

A_1 Agricultural production per hec. of net area sown (tones)

A_2 Consumption of power per hec. in Agriculture(Kwh.)

A_3 Percentage of net irrigated area to net area sown

A_4 Consumption of fertilizerper hec. of gross cropped area (kg)

A_5 Cropping intensity (%).

District wise composite indices of agricultural development indicators

Districts	Z_1	Z_2
	1992-93	*2004-05*
Kurukshetra	1.64948	1.90773
Bhiwani	-0.62589	-1.68465
M.Garh	-0.94631	-1.24874
Yamuna Nagar	-0.55143	-.97345
Panipat	0.58426	0.12274
Gurgawn	-0.71324	0.54228
Ambala	0.54092	0.65585
Kethal	2.99549	3.18306
Jind	-0.09814	-0.80726
Sonopat	-0.75775	-0.04585
Karnal	1.62036	1.70153
Faridabad	-1.20422	-1.89113
Hissar	0.98888	1.181769
Sirsa	-0.69889	0.604918
Rewari	-0.77057	-0.18907
Rohatak	-0.66254	-1.82793
Total	03.4329	05.109677
Mean	0.419726923	0.47656228
S.D.	1.910344616	2.010702331
C.V	434.43477	460.57816

Growth and Service Delivery: Building on India's Success. The study outlined two most pressing challenges for public action in India are: (i) Institutional reform to enhance the capacity of public sector institutions to ensure the effective delivery of core services and (ii) sustaining rapid growth and making the process of economic growth more inclusive–across sectors, across regions, and bringing the benefits of higher income and living standards to more people.

CONVERGENCE: THEORETICAL DEVELOPMENTS

The most striking aspect of the per capita income distribution of the economies is very large and has persistent disparities. There have been "growth disasters" economies in which per capital income have fallen and "growth miracles" economies which have risen rapidly up the world income distribution. One of the first questions we might ask is whether there is a systematic tendency for poor countries to grow faster than rich ones, which would allow them to catch up.

Convergence theory dates back to the standard neo-classical growth model with exogenous technological process and a closed economy. Neo-classical growth models for closed economies are given by Solow (1956), Cass (1965) and Koopmans (1965). Their model states that the per capita growth rate tends to be inversely related to the starting level of output or income per person in the economy. In particular, if economies are similar in respect to preferences and technologies then poor economies grow faster than rich ones.

The seminal theoretical contributions of Romer (1986) and Lucas (1988) on endogenous growth mechanisms have generated a lively empirical literature. This literature has primarily focused on the so called 'convergence hypothesis'. The hypothesis asserts that differences in contemporaneous per capita income between any pair of economies will be transitory so long as the two economies possess identical technologies, preferences and population growth rates [Bernard and Darlauf, (1996)]. By implication the effects of initial stocks of human and physical capital on per capita output differences will be transitory so long as these other factors are accounted for.

In recent years, since the appearance of the pioneering studies [Baumol (1986) Barro (1991) and Mankiew *et al.* (1992)] considerable effort have been devoted to identifying changes over time in the cross sectional distribution of per capita income between economies. These studies have observed that economies with low initial income tend to grow faster than economies with high initial incomes with controlled savings rate and population growth. This finding has been treated as an evidence of convergence. There are many studies of the international processes of growth and convergence across countries [Baumol (1986), De Long (1988), Dowrick and Nguyen (1989) and Barro and Sala-i-Martin (1995) among others] and regional growth patterns within any given country.

A summary statistics is used to measure convergence directly employing coefficient of variation (C.V.) by Baumol and Wolff (1988), Metha, Crihfield and Gietz (1992), Izraeli and Murphy (1997) and Roberts (1997). Izraeli and Murphy (1997) have also used Bourguignon's L-statistics to measure convergence directly.

Empirical models which estimate a relationship of the form $y_{it} = f(y_{io})$, where 'y_{it}' measures the size of some economic unit 'i' at time 't', have a long tradition in the economics literature. It is known that cross sectional estimation is hazardous in the case where 'y_{it}' tends to converge towards steady state values, which differ across 'i'; or in other words, when there are heterogeneous individual effects. Islam (1995) for example, has pointed out that, since the omitted individual effects are correlated with the explanatory variable 'y_{io}', the estimated coefficient on the latter is upward biased. Recent studies like Funke and Strulik (1999) and Goddard and Wilson (2001) have used panel estimation of convergence.

EVIDENCE OF CONVERGENCE/DIVERGENCE IN INDIA

There are a number of studies on the convergence literature across Indian states. Cashin and Sahay (1996) Marjit and Mitra (1996), Ghose, Marjit and Neogi (1998), Rao, Shand and Kalirajan (1999) and Dasgupta, Maiti, Mukherjee, Sarkar,

Chakrabarti (2000) have addressed the issues relating to convergence in India. Cashin and Sahay (1996) examined the growth experience of 20 states of India during 1961-91 using Solow-Swan neoclassical growth model in a cross-sectional framework and found evidence of absolute convergence. They also verified that central grants speed up the convergence process. Marjit and Mitra (1996) have stated that the states of India are diverging in terms of their per capita income. They used scattered plots between base period per capita net state domestic product and the annual average growth rates of the states to identify 'divergence'. Ghose, Marjit and Neogi (1998) obtained a relationship between per capita net state domestic product and its growth rate across the states over 35 years to conclude 'divergence'. Rao, Shand and Kalirajan (1999) have shown divergence in the per capita income across states during 1960-61 to 1994-95. They have observed that inter-state disparities are accentuating over the years. Dasgupta et al. (2000) offer analytical description of the economic performance of Indian states using per capita (net) state domestic product over the year 1960-61 to 1995-96. Their analysis draws the conclusion that Indian states tend to diverge in per capita SDP but converge in shares of different sectors in the SDP.

However, all of the studies have taken per capita (net) state domestic product as the source of convergence or divergence using cross-sectional data. Again, none of the studies have adopted panel data technique. Alternatively, growing convergence literature uses panel technique. This technique seems reasonable since convergence is by definition a dynamic concept, which cannot be captured by cross-sectional studies. In view of this the objective of the present study is to verify the convergence principle via per capita income using panel data estimates. Besides we also analyse the convergence principle for the per capita agricultural income of the states. This is because agriculture is the lagging sector of the Indian economy.

ANALYTICAL FRAMEWORK

State level data of India on the per capita income are taken as the source of convergence in the present analysis. Per capita incomes for the fourteen major states of India at constant prices are collected from the Central Statistical Organisation. Also the per capita agriculture income is taken for the fourteen major states and the convergence principle is tested for it. The availability of data restricted the period of study from 1980 to 2004 for the fourteen major states of India.

Three distinct methods have been adopted in the present study to measure convergence.

Direct Measures of Convergence

Direct measures of convergence seek to measure convergence directly using a summary statistic such as the square of coefficient of variation (C.V.) or Bourguignon L-statistics (1979). Both the measures are aggregative, income zero homogeneity, obey Pigou-Dalton condition and also are additively decomposable. These statistics are not sensitive to the national rate of inflation, as both the measures are proportions (pure-numbers).

$$\text{Here, C.V.} = \frac{\sigma}{\overline{X}} \times 100 \qquad (1)$$

And 'σ' represents the standard deviation in the per capita income of the fourteen major states at a particular time period and $\overline{X}$ the simple arithmetic mean.

Likewise,

$$L = \log\left(\frac{1}{n}\sum_{i=1}^{n} yi\right) - \frac{1}{n}\sum_{i=1}^{n} \log yi, \qquad (2)$$

Here 'y_i' is the per capita income of the 'i' th state.

Indirect Measurement of Convergence

Indirect measurement of convergence deals with two concepts of convergence. They are (i) b-convergence and (ii) s-convergence. Absolute b-convergence occurs if poor economies tend to grow faster than rich ones. Let 'y_{it}' be the per capita income for a cross section of economies (of states) in time period 't'.

$$\text{Also let, } \gamma_{i,t,t+T} = [\log (y_{i,t+T} / y_{i,t})] \quad (3)$$

be the 'i' th economy's growth rate of per capita income between 't' and 't+T' and let $\log(y_{i,t})$ be the logarithm of economy i's income per capita at time 't'.

$$\text{If, } g_{i,t,t+T} = \alpha + \beta \, . \log (y_{i,t}) + \epsilon_i \quad (4)$$

$$\text{or, } \gamma_{i,t,t+T} = \alpha + [\log (y_{i,t}) \, . \, (1-e^{-bT})] + \epsilon_i \quad (5)$$

The above equation is estimated and if $\beta<0$, then the data set exhibits β-convergence. Here, $\beta = (1-e^{-bT})$, where, 'σ' is the speed of convergence.

The concept of σ-convergence can be defined as, a group of economies are converging in the sense of 'σ' if, the dispersion of per capita income tends to decrease overtime. That is, $\sigma_t > \sigma_{t+T}$ where 'σ_t' is the time't' standard deviation of $\log (y_{i,t})$ across 'i'.

Panel Data Estimates

One common issue that arises in panel data estimation is whether the individual effects are to be thought of as 'fixed' or 'random'. In the later case the effects are to be uncorrelated with the exogenous variables included in the model. It is clear that estimators relying on such assumptions [for example, GLS in Maddala (1971)] are not suitable because it is precisely the fact of correlation that forms the basis for panel approach. The least square with dummy variable estimator, which is based on the fixed effect assumption, is still permissible [Islam (1995)].

By exploiting a pooled time-series and cross-sectional structure, less information is lost than taking an average value for each variable for each region over the period in question. In other words, panel data estimates introduced here provide more transparent and intuitively more reasonable results [Funke and Strulik (1999)]. The growth equation as a dynamic panel data model is formulated below.

Let 'y_{it}' be the ratio of (Y_{it}/Y_t^*) where 'Y_{it}' is the 'i' th state per capita income in period 't' and Y_t^* is aggregate (average) per capita income in fourteen major states of India in period 't'. Allowing for different steady states and different convergence rates across the 14 major states gives the following expression for 'y_{it}'.

The model here is

$$\ln (y_{it}) = \gamma . y_{i,o} + \rho_i . \ln (y_{i,t-1}) \quad (6)$$

In the above equation, i = 1,2, ...N and t = 1,2.... T.

The model allows for initial per capita incomes to influence future growth through the parameter 'γ'. The corresponding statistical model can be written as

$$\ln (y_{it}) = \alpha_i + \rho_i . \ln (y_{i,t-1}) + \epsilon_{it} \quad (7)$$

Note that β_i = -ln (ρ_i) and 'ϵ_{it}' represent independently drawings from a normal distribution with mean zero and variance σ^2_ϵ. The model allows the initial conditions to influence future growth through α_i. The (relative) steady state value of w_{it} is (α_i/1-ρ_i) which is a measure of efficiency and reflects total factor productivity in region 'i'.

AGRICULTURAL GROWTH & STATE INCOME

The empirical findings of the study are presented below.

Direct Measures of Convergence

In the direct measures of convergence two statistics, i.e. the square of coefficient of variation and Bourguignon L-statistics are used to measure the degree of variation in the per capita state income and per capita agricultural income. The estimates of the two statistics for the fourteen major states of India during 1980 to 2004 are presented in Table 1.

Both the statistics exhibit similar behaviour over the period of study for both the variables. For the fourteen major states both the square of coefficient of variation and L-statistics have first increased and then decreased. The square of coefficient of variation has increased from 824.17 in 1980-81 to 1332.94 in 1996-97 for the per capita income. The square of coefficient of variation was 1288.82 in the year 1997-98 then it increased continuously to 1355.60 in the year 2004. Similarly the L-statistics for per capita income has increased from 0.0169 in

Table 1
Square of C.V. and L Statistics of Per Capita State Income and Per Capita State Agricultural Income

	Per Capita Income		*Per Capita Agriculture Income*	
Year	*Sq of C.V.*	*L Statistics*	*Sq of C.V.*	*L Statistics*
1980	824.177865	0.01691118	1438.183801	0.023425899
1981	886.433943	0.01795177	1530.040043	0.025956931
1982	940.834593	0.01889752	1828.284025	0.029754227
1983	815.433479	0.01657738	1338.109155	0.023740213
1984	912.287196	0.01782939	1580.066876	0.024723327
1985	973.011135	0.01826326	2049.125619	0.029494983
1986	992.652649	0.01915199	2054.269276	0.030123151
1987	1050.76892	0.02006621	2365.525322	0.033498083
1988	996.638239	0.0197682	1897.816313	0.029441169
1989	1079.57884	0.0217668	2087.541978	0.02985982
1990	1095.52289	0.02255402	2171.834764	0.03155016
1991	1115.63534	0.02302166	2545.786651	0.034875083
1992	1262.45321	0.0278001	2303.308367	0.034381963
1993	1275.15333	0.02856558	2257.527929	0.03237548
1994	1220.47057	0.02795486	1986.166214	0.029406835
1995	1319.57853	0.03250656	2061.980948	0.03173324
1996	1332.94094	0.03152704	2049.569572	0.03280999
1997	1288.8239	0.03186358	1871.621875	0.030565787
1998	1282.82503	0.03258813	1700.280284	0.027665004
1999	1270.13553	0.03241464	2172.459108	0.033766652
2000	1250.4772	0.03114107	2610.758033	0.039363937
2001	1264.06532	0.03341442	2230.971365	0.034335106
2002	1357.34244	0.03410833	2793.234038	0.040233263
2003	1333.97535	0.03603757	2512.239079	0.040806554
2004	1355.60395	0.03595842	2492.755906	0.037078535

1980-81 to 0.03595 in 2004-05. Both the statistics measure the variability of per capita income across the fourteen major states. Thus they indicate that there is divergence in the per capita income for the fourteen major states.

In case of agricultural income although the square of coefficient of variation do not indicate any constant increase still it increased with ups and downs. But the L-statistics increased almost continuously. Thus there is divergence in the per capita agricultural income as indicated by the L-Statistics.

The variability among the states in the per capita income and per capita agricultural income during the years 1980-81 to 2004-05 increased. This indicates that the fourteen major states have diverged in their per capita income.

Indirect Measurement of Convergence: Testing the Catch up Hypothesis

The method used to test the β-convergence (or catch up hypothesis) has been discussed earlier. The functional form of the equation as log-log model and the dependent variable is defined as the rate of growth of state per capita income ($y_{i,t}$) and per capita agricultural income over the time period. The estimated parameters of convergence are presented in Table 2.

Table 2
Cross-Sectional Regression of Per Capita State Income
[(dépendent variable $log\ (y_{2004}/y_{1980})$]

Parameters	*Ordinary Least Square*
Constant	-1.1436
	(-1.5505)
log (y_{1980})	0.3977**
	(2.0035)
R^2	0.2506
σ-convergence in the per capita state income	
σ_{1980}	σ_{2004}
0.1273	0.1983

Note: Figures in the parentheses indicates the respective t-value.
** indicates significance of the t-value at 1% level of significance.

The estimated value of 'β' in case of per capita income during 1980-81 to 2004-05 for the fourteen major states is 0.3977. Clearly, β>0 and hence over the period 1980-81 to 2004-05 we have β-divergence in the per capita income. Here, σ_t is 0.1273 but σ_{t+T} is 0.1983 which indicates there is σ-divergence. Thus, both b and σ-convergence indicate divergence in the state per capita income for the period 1980-81 to 2004-05.

Likewise the estimated value of σ_t in case of per capita agricultural income is 0.1409 during 1980-81 but σ_{t+T} is 0.17403 during 2004-05 which is presented in Table 3. As the value of σ increased over time thus, there is s-divergence. On the other hand the estimated value of 'β' is 0.0253. Thus β>0 and we have β divergence for the per capita agriculture income during the same period.

Table 3
Cross-Sectional Regression of Per Capita Agricultural Income [(dependent variable log (y^a_{2004}/y^a_{1980}))]

Parameters	*Ordinary Least Square*
Constant	-0.0142
	(-0.0215)
log (y^a_{1980})	0.0253
	(0.1274)
R^2	0.0135
σ-convergence in the per capita state agricultural income	
σ_{1980}	σ_{2004}
0.1409	0.1740

Note: Figures in the parentheses indicates the respective t-value.

Panel Data Estimates

(a) Per capita state income: The set of results obtained through the panel data estimates for the fourteen major states of per capita income are presented in Table 4. The individual per capita income varies across the fourteen major states. The divergence rate varies from a low 0.9 per cent in Bihar to a maximum of 102 per cent in Gujarat. The ordering of the steady states for the states are Punjab, Maharastra, Haryana, Gujarat, T.N., Karnataka, U.P. and Kerala. Divergence across developing countries carries over to the question of regional divergence. However, states like Andhra Pradesh, Bihar, Madhya Pradesh, Orissa, Rajasthan and W.B. have negative steady state value. The steady state value can be negative, implying in the steady state, these states grow at a lower rate than the average. The significant 'a_i' coefficients in five states indicate persistence of inequality, with the passage of time some rich regions tend to remain rich while other poor regions continue to lag behind.

Pooling estimates of convergence for different categories of states (HIS, MIS and LIS) are calculated and presented in Table 6. Here, the divergence rate is maximum in the high income category of states (20.69%) followed by the divergence rate in the middle (8.38%) and low income states (1.83%). The divergence rate is 10.08%, when data of all the fourteen major

Table 4
Panel Data Estimates of Per Capita Income of Fourteen Major States

State	α	ρ	*Divergence Rate*	*Steady State*
Andhra	-0.017795	0.645558*		
	(-1.21)	(3.98)	0.43764	-0.05021
Bihar	-0.096907	0.911433*	0.092737	-1.09417
	(1.10)	(9.19)		
Gujarat	0.166760*	0.357146	1.029611	0.259406
	(3.18)	(1.78)		
Haryana	0.142290*	0.580690*	0.543538	0.339343
	(2.39)	(3.34)		
Karnataka	0.008518	0.869978*	0.139287	0.065512
	(0.90)	(7.26)		
Kerala	0.000496	0.810970*	0.209524	0.002624
	(0.06)	(5.87)		
M. P.	-0.063267*	0.756818*	0.278632	-0.26016
	(-1.98)	(5.26)		
Maharashtra	0.070772	0.800803*	0.22214	0.355286
	(1.63)	(6.26)		
Orissa	-0.111115*	0.772060*(	0.258693	-0.48747
	(-1.99)	6.32)		
Punjab	0.019909	0.944647*	0.056944	0.359673
	(0.36)	(7.96)		
Rajasthan	-0.198218*	0.746898	0.291827	-0.78315
	(-5.12)	(0.70)		
T.N.	0.024866	0.820017*	0.19843	0.138157
	(1.85)	(7.83)		
U.P.	-0.001524	1.033208*	-0.03267	0.045893
	(-0.07)	(21.53)		
W.B.	-0.008079	0.889705*	0.116865	-0.07325
	(-0.50)	(6.95)		

Note: *indicates significance of the t-value at 1% level of significance.

states are pooled. However, the steady state value of only the low Income States is negative (-1.07). These imply that in steady state these regions grow at a lower rate than the average.

(b) Per capita state agricultural income: The set of results obtained through the panel data estimates for the fourteen major states of per capita agricultural income are presented in Table 5. The individual per capita agricultural income varies across the fourteen major states. The divergence rate varies from a high 205.89 per cent in Gujarat to a minimum of 16.48 per cent in W.B. The ordering of the steady states for the states are Punjab, Haryana and Gujarat. Divergence across developing

countries carries over to the question of regional divergence. States like Andhra Pradesh, Bihar, Karnataka, Kerala, Madhya Pradesh, Maharashtra, Orissa, T.N., Rajasthan, U.P. and W.B. have negative steady state value. The steady state value can be negative, implying in the steady state, these states grow at a lower rate than the average. The significant 'a_i' coefficients in ten states indicate persistence of inequality, with the passage of time some rich regions tend to remain rich while other poor regions continue to lag behind.

Table 5
Panel Data Estimates of Per Capita Agricultural Income of Fourteen Major States

State	α	ρ	*Divergence Rate*	*Steady State*
Andhra	-0.031710	0.430067*	0.843814	-0.05564
	(-1.46)	(2.13)		
Bihar	-0.168416*	0.650176*	0.430512	-0.48143
	(-2.25)	(4.19)		
Gujarat	0.010529	0.127590	2.058933	0.012069
	(0.21)	(0.61)		
Haryana	0.438430*	0.190899	1.656011	0.541873
	(3.85)	(0.91)		
Karnataka	-0.010520	0.368598	0.998049	-0.01666
	(-0.48)	(1.85)		
Kerala	-0.108496*	0.659375*	0.416463	-0.31852
	(-2.01)	(3.79)		
M. P.	-0.103972*	0.503932*	0.685314	-0.20959
	(-3.85)	(2.48)		
Maharashtra	-0.165932*	0.315438	1.153793	-0.24239
	(-3.08)	(1.54)		
Orissa	-0.114703*	0.679122*	0.386954	-0.35747
	(-2.09)	(4.57)		
Punjab	0.384779*	0.528310*	0.638072	0.815746
	(3.01)	(3.30)		
Rajasthan	-0.112625*	0.617933	0.481375	-0.29478
	(-3.02)	(0.55)		
T.N.	-0.172720*	0.455322*	0.78675	-0.3171
	(-2.75)	(2.52)		
U.P.	-0.139756*	0.414574*	0.880504	-0.23873
	(3.17)	(2.22)		
W.B.	-0.025547	0.848047*	0.164819	-0.16812
	(-0.72)	(7.37)		

Note: * indicates significance of the t-value at 1% level of significance.

Table 6
Pooling Estimates of Per Capita Income

State	α	ρ	Divergence Rate	Steady State
FMS	-0.001586	0.904107*	0.100808	-0.01654
	(-0.47)	(107.85)		
HIS	0.065669*	0.813094*	0.206909	0.351348
	(3.28)	(14.52)		
MIS	0.004316	0.919584*	0.083834	0.053671
	(1.11)	(23.69)		
LIS	-0.019676	0.981770*	0.018398	-1.07932
	(-1.52)	(38.45)		

Note: FMS-Fourteen Major States, HIS - High Income States, MIS -Middle Income States and LIS -Low Income States. * -Indicates significance of t-value at 1% level of significance.

Table 7
Pooling Estimates of Per Capita Agricultural Income

State	α	ρ	Divergence Rate	Steady State
FMS	-0.006689	0.925627*	0.077284	-0.08994
	(-0.83)	(42.30)		
HIS	0.020363	0.928995*	0.073652	0.286783
	(0.97)	(22.84)		
MIS	-0.031449	0.819998*	0.198453	-0.17471
	(-2.29)	(15.50)		
LIS	-0.074392*	0.711859	0.339875	-0.25818
	(-3.63)	*(10.69)		

Note: FMS-Fourteen Major States, HIS-High Income States, MIS-Middle Income States and LIS -Low Income States. * -Indicates significance of t-value at 1% level of significance.

Pooling estimates of convergence for different categories of states (HIS, MIS and LIS) are calculated and presented in Table 7. Here, the divergence rate is maximum in the low income category of states (33.98%) followed by the divergence rate in the middle (19.84%) and low income states (7.36%). The divergence rate is 7.72%, when data of all the fourteen major states are pooled. However, the steady state value of only the low Income States is negative (-1.07). These imply that in steady state these regions grow at a lower rate than the average.

CONCLUSION

The liberalization policy of 1991 has allowed the states to play a larger role in determining their development paths and attracting investment. India's success in fostering economic growth has created new challenges. The question now is how India can sustain that success and expand the opportunities for all to participate in it and benefit from it. Increased growth since the mid 1980s has helped to substantially reduce the national poverty rate to 22% of the population in 2004, with the speed of poverty reduction appearing to increase between 1999 and 2004. To meet one of its Millennium Development Goals, of halving poverty by 2015, the government is aiming to achieve an even higher medium term economic annual growth rate of 10%. In addition growth needs to become more inclusive by increasing the prosperity of poorer states, whose economies have expanded at a slower pace than those of richer states in the past decade, and so reducing their difficulties in lowering poverty.

Important challenges persist for agriculture. The need to accelerate growth in India's agricultural sector prompts special consideration due to the slow down on agricultural production, employment and wages. Where growth in nonagricultural sector has accelerated, the reallocation of labour out of agriculture is lagging, concentrating poverty in rural areas and widening rural-urban income disparities. More divergence of agricultural income in low income states implies the need for strategies for these regions. There should be new strategy which should rely much more on policies to liberalise agricultural inputs and outputs by reforming land administration and tenancy requirements and by supporting research and development that is tailored by specific needs. A workable strategy to accelerate productivity in agriculture has to promote four directions for agriculture in the overall rural economy. They are:

Intensification: Increase the productivity of traditional crops through greater use of irrigation, high yield verities, agrochemical inputs and mechanization.

Diversification: Shift to new more profitable crops (fruits, vegetables, higher value cereals, medicinal plants) and livestock.

Non farm linkage: Emphasize related activities that add value to agricultural products such as trading, agro processing and supplying inputs. *Exit*: Switch from agriculture to other more productive activities.

The government of India's justifiable concern with the inclusiveness of economic growth can be addressed by focusing on expanding the regional scope of economic growth, expanding access to assets and thriving markets and expanding equity in the opportunities for the next generation of Indian citizens.

NOTES

1. High income states (HIS) are Gujrat, Maharartra, Punjab and Haryana, Middle income states (MIS) are Tamil Nadu, Karnataka, Kerala, West Bengal and Andhra Pradesh and low income states (LIS) are Uttar Pradesh, Rajasthan, Madhya Pradesh, Bihar and Orissa.
2. The concept and name were first presented in 1989–1990 by John Williamson, an economist from the Institute for International Economics, Washington DC. The Washington Consensus contained a set of 10 recommendations that synthesized the policies considered necessary for the recovery of Latin American economies from the financial crises of the 1980s and cycles of high inflation and low growth.
3. Sixteen leading economists, gathered in Barcelona in September 2004 to discuss effects of economic reforms, lessons for future policy making, and performance of international economic systems, and issued a consensus on growth and development.

REFERENCES

Acharya S. (2007), *Can India Grow without Bharat?* Academic Foundation, New Delhi.

Aghion P. and Howitt P. (2005), 'Appropriate Growth Policy: A Unifying Framework', Harvard University, Working Paper.

Aghion, P. and Williamson, J. G. (1998), *Growth, Inequality and Globalization: Theory, History and Policy.* Cambridge: Cambridge University Press.

Alan Winters L. and Yusuf S. (Ed.) (2007), *Dancing with Giants: China, India, and the Global Economy*, the World Bank and the Institute of Policy Studies, Singapore.

Barro, R. J. (1991), 'Economic Growth in a cross-section of Countries', *Quarterly Journal of Economics*, 106: 407-444.

Barro, R. J. and Sala-i-Martin, X. (1995), *Economic Growth*, Singapore: McGraw-Hill International Editions.

Barro, R. J. and Sala-i-Martin, X. (1992), 'Convergence', *Journal of Political Economy*, 100 (May), 223-252.

Baumol, W. J. (1986), 'Productivity Growth, Convergence and Welfare: What the Long-run Data Show', *American Economic Review*, 76 (5): 1072-1085.

Baumol, W. J. and Wolff, E. N. (1988), 'Productivity Growth, Convergence and Welfare: Reply', *American Economic Review*, 78 (5): 1155-1159.

Bloom D. E., Canning D., Hu L., Liu Y., Mahal A. and Yip W. (2006), 'Why has Chinas Economy Taken Off Faster than India's?', Princeton University.

Bosworth B. and Collins S. M. and Virmani, A. (2007), 'Sources of Growth in the Indian Economy', NBER Working Paper 12901, Cambridge, MA, February.

Bosworth B. and Collins S.M. (2007), 'Accounting for Growth: Comparing China and India', NBER Working Paper No. 12943, Cambridge, MA, February.

Bourguignon, F. (1979), 'Decomposable Income Inequality Measures', *Econometrica*, 47(4): 901-920.

Broadman H. G. (2007), *Africa's Silk Road: China and India's New Economic Frontier*, World Bank, Washington, DC.

Cashin, P. and Sahay, R. (1996), 'Regional Economic Growth and Convergence in India', *Finance and Development*, March, 49-52.

Das, T. K. (2000), 'Economic Growth and Convergence across India States', International Symposium on Development Policies for the New Millenium, IGIDR, Mumbai, 12-14 July, 89-94.

Dasgupta, D. et al. (2000), 'Growth and Inter State Disparities in India', *Economic and Political Weekly*, July 1st, 35(27): 2413-2422.

De Long, J. B. (1988), 'Productivity Growth, Convergence and Welfare: Comment', *American Economic Review*, 78 (5): 1138-1154.

Dowrick, S. and Nguyen, D. T. (1988), 'OECD Comparative Economic Growth 1950-1985: Catch up and Convergence', *American Economic Review*, 79: 1010-1030.

Ernst & Young (2006), *Doing Business in India*, Ernst & Young Private Limited, New Delhi.

Funke, M. and Strulik, H. (1999), 'Regional Growth in West Germany: Convergence Divergence', *Economic Modeling*, 16: 489-502.

Ghosh, B., Marjit, S. and Neogi, C. (1998), 'Economic Growth and Regional Divergence in India, 1960 to 1995', *Economic and Political Weekly*, June 27: 1623-1630.

Goddar, J. and Wilson, J. (2001) 'Cross Sectional and Panel Estimation of Convergence', *Economic letters*, 70: 327-333.

Gordon, J. and Gupta, P. (2003), 'Understanding India's Services Revolution', IMF-NCAER Conference, New Delhi, November 14-16.

Government of India (2006), *Towards Faster and More Inclusive Growth An Approach to the 11th Five Year Plan (2007-2012)*, Planning Commission, Yojana Bhavan, New Delhi, December.

Hausmann R., Pritchett L., and Rodrik D. (2004), 'Growth Accelerations', Harvard University, May.

Hausmann R., Rodrik D. and Velasco A. (2005), 'Growth Diagnostics', John F. Kennedy School of Government, Harvard University, Cambridge, MA, March.

IMF (2006), 'Understanding Growth', *Finance & Development*, Vol. 43, No. 1.

IMF, (2006), *Learning From Each Other: Reforms and Policies for Sustained Growth*, Washington, DC.

Islam, N. (1995), 'Growth Empirics: a panel data approach', *Quarterly Journal of Economics*, 60: 1127-1170.

Izraeli, O. and Murphy, K. (1997), 'Convergence in State Nominal and Real per capita Income', *Public Finance Review*, 25 (6): 555-576.

JM Morgan Stanley (2006), *India and China: New Tigers of Asia, Part-II*, Special Economic Analysis, Mumbai, June.

Khanna T. (2004), 'India's Entrepreneurial Advantage', *The McKinsey Quarterly*, Special Edition, China Today, P. 111.

Lane P.R. and Schmukler S.L. (2007), The International Financial Integration of China and India, WPS 4132 World Bank, Policy Research Working Paper.

Leipziger D. and Zagha R. (2006), 'Getting Out of the Rut', *Finance & Development,* Vol. 43, No. 1, p. 16-18.

Lin, J. Y. (2004), 'Development Strategies for Inclusive Growth in Developing Asia', China Center for Economic Research, Working Paper No. E2004007.

Lucas, R. E. (1988), 'On the Mechanics of Economic Development', *Journal of Monetary Economics,* 22 (July), 3-22.

Lucas, R. E. (1990), 'Why Does not Capital Flow from Rich to Poor Countries?,' *American Economic Review,* 80(May), 92-96.

Maddala, G.S. (1971), 'The use of Variance Components Model in Pooling Cross-section and Time Series Data,' *Econometrica,* 39: 341-358.

Mankiew, N. G., Romer, D. and Weil, D. N. (1992), 'A Contribution to the Empirics of Economic Growth', *Quarterly Journal of Economics,* 107 (May), 407-437.

Marjit, S. and Mitra, S. (1996), 'Convergence in Regional Growth Rates: Indian Research Agenda', *Economic and Political Weekly,* August 17: 2239-2242.

Metha, S. J., Crihfield, J. B and Giertz, F. (1992), 'Economic Growth in American States: The End of Convergence', Working Paper No. 20, The Institute of Governments and Public Affair, University of Illinois.

Oxley, L. and Greasley, D. C. (1995), 'A Time-Series Perspective on Convergence: Australia, UK and USA since 1870', *The Economic Record,* 71(214), (September), 259-270.

Poirson, H. (2007), India: Country Study, IMF Research Bulletin, June.

Purfield C. and Schiff J. (Ed) (2006), *India Goes Global: Its Expanding Role in the World Economy,* International Monetary Fund, Washington, DC.

Rao, M. G., Shand, R. T. and Kalirajan, K. P. (1999), 'Convergence of Incomes across Indian States', *Economic and Political Weekly,* March 27: 769-778.

Roach, S. (2006), *India and China: New Tigers of Asia, Part-II,* JM Morgan Stanley Research, Mumbai, June.

Roberts, C.A. (1997), 'Interregional per capita Income Differences and Convergence: 1880-1950', *Journal of Economic History,* 39 (1): 101 112.

Romer, P. (1986), 'Increasing Returns and Long-run Growth', *Journal of Political Economy,* 94 (October), 1002-1037.

Romer, P. (1989), 'Capital Accumulation in the Theory of Long-run Growth', in R. J. Barro (ed.) *Modern Business Cycle Theory,* Cambridge-Mass: Harvard University Press.

Solow, R. M. (1956), 'A Contribution to the Theory of Economic Growth', *Quarterly Journal of Economics,* 70 (February), 65-94.

United Nations (2006), *World Economic and Social Survey: Diverging Growth and Development,* Economic & Social Affairs.

Wilson D. and Purushothaman R. (2003), *Dreaming with BRICs: The Path to 2050,* Goldman Sachs Global Economics, Paper No. 99, New York.

World Bank (2005), *Economic Growth in the 1990s: Learning from a Decade of Reforms,* Washington, DC.

World Bank (2006), *Inclusive Growth and Service Delivery: Building on India's Success,* Development Policy Review, Washington, DC.

World Bank, Data on Poverty in India, Washington, D.C.

Zagha R., Nankani G. and Gill I. (2006), 'Rethinking Growth', *Finance & Development,* Vol. 43, No. 1, p. 7-11.

5

AGRICULTURAL DISPARITY ACROSS INDIAN STATES: DETERMINANTS AND POLICY IMPLICATIONS

BIMAL KISHORE SAHOO

INTRODUCTION

Notwithstanding, it is a clinch, it is true that agriculture is the backbone of Indian economy. In India two-thirds of the labour force engaged in agriculture and related activities.Given that masses depend upon agriculture for their live-hood, agriculture has always been in the forefront of policy discussions in India. During the first two decades of planning era, the process of agricultural development was commenced through land reforms, public investments in creating irrigation, power and road infrastructure, setting up State Agricultural Universities for agricultural education, research and extension services, opening bank braches in small towns and rural areas for providing institutional credit to the farmers for purchasing modern technology, inputs, farm machines and electric and diesel pumps for wells and tube-wells. The Green Revolution (GR) approach that was started in the late 1960s, has transformed the Indian economy from a food-deficit to a self-sufficient one. But unfortunately, its benefits were spread disproportionately in favour of the wealthier farmers, especially of agriculturally developed regions, avoiding the small and marginal farmers, in particular and agriculturally underdeveloped regions in general. Although, the GR technology was size-neutral; it was not resource-neutral. Therefore, its benefits have largely been reaped by the large farmers of developed regions as they have better access to resources. There have, now, been ample evidences that inter-region and inter-farm disparities in the agricultural development has widened in the post-GR period.

Although, economic reforms measures initiated in the Indian economy in 1991 did not directly target the agriculture; they did have indirect impact on it via reduction in the developmental role of the government and its capacity to generate resources for investment. A reduction in custom and excise duties owing to the process of tax liberalisation has reduced the tax-GDP ratio from10.1 per cent in 1990-91 to 8.8 per cent in 2002-03. Due to fall in tax-GDP ratio and a rise in the non-plan expenditure, revenue deficit as percentage of the fiscal deficit increased from 49.4 per cent in 1990-91 to 82.2 per cent in 2002-03. These policies have also affected the State governments' fiscal resources as net transfer of funds from the Central Government to the State Governments has declined. Agriculture, being a State subject, and with dwindling State finances, the public investment in it has been adversely affected (Singh, 2005). Further, benefits of economic reforms did not evenly distributed across Indian States due to reasons such as, differences in the quality of infrastructure, governance, human capital, and the location of States. The States located in the coastal regions have performed relatively better in terms of higher private investment, including foreign direct investment and achieved higher growth in the Net State Domestic Product (NSDP) which enhanced the state governments' revenues and their capacity to invest in socio-economic infrastructure, including irrigation, power, road and agriculture, whereas States located in the North and North-eastern regions could not attract more private domestic as well as foreign direct investment. As a result, disparity in economic development across regions has increased. Several studies have been conducted on the subject of inter-state disparity in India. However, most of these studies are based on aggregate data of NSDP. These studies have conclusive evidence that in the postreform period, economic disparity among states has increased (Kurian, 2000; Singh *et al,* 2003; Bhattacharya and Sakthivel, 2004; Das, 2002; Shetty, 2003). But whether disparity is increased due to differences in agricultural or non-agricultural NSDP has not yet been given due emphasis. Keeping this in backdrop, this paper examines the inter-state disparity in the agricultural productivity; identifies the major factors in explaining this disparity and suggests policy interventions to alleviate it.

DATA SOURCES AND VARIABLES

The study is mainly based on the secondary data collected from various published sources for the years 1981 to 2003. The main data sources are official websites of Ministry of Agriculture, Ministry of Statistics and Programme Implementation and Reserve Bank of India (RBI), Data on agricultural inputs have been compiled from the Database of National Centre for Agricultural Economics and Policy Research (NCAP). Fifteen major Indian States, which together constitute about 90 per cent of total population of the country, are considered for studying the inter-state disparity in the agricultural performance.

Besides estimating annual compound growth rates, simple measure of coefficient of variation (COV) and Gini Coefficient has been calculated to know the variation across States. Regression analysis has been conducted to identify the factors responsible for the inter-state disparity. Some farm and non-farm related variables such as, fertilizer use, cropping intensity, gross irrigated area, farm mechanization and length of pucca roads, etc. have been considered as explanatory variables to explain the disparity across States.

TRENDS IN NET STATE DOMESTIC PRODUCT

In order to know the performance of individual States during the last two decades, Per Capita Net State Domestic Product (PNSDP) is considered as a proxy variable for the per capita income. Table 1 provides information on the PNSDP at factor cost for Triennium Ending (TE) 1980-82, 1990-92 and 2001-03. It is evident from the table that there exists a wide disparity in the PNSDP across States. In TE 1980-82, States like Assam, Bihar, Uttar Pradesh (UP), Madhya Pradesh (MP), Rajasthan and Orissa had PNSDP just less than half that of Punjab. As against PNSDP of Rs.2818 in Punjab, the corresponding figure for Bihar was only Rs.933. The table also shows that the gap between poorer and richer States has grown over the years. For example, PNSDP in Punjab was 4.17 times that of Bihar in TE 2001-03 as against 3.02 times in TE 1980-82. The values of coefficient of variation in the PNSDP across States show that economic disparity among the States has increased during the last two decades.

Table 1
State-wise Per Capita Net State Domestic Product at Constant Prices

	1980-82		*1990-92*		*2001-2003*	
States	*PNSDP (Rs.)*	*Rank*	*PNSDP (Rs.)*	*Rank*	*PNSDP (Rs.)*	*Rank*
AP	1504.00	8	2077.67	8	1 0999.67	8
Assam	1374.33	10	1558.67	13	6251.00	12
Bihar*	933.00	15	1106.33	15	37] 9.67	15
Gujarat	2010.67	4	2704.33	4	14850.00	4
Haryana	2418.67	3	3476.33	3	14875.67	3
Karnataka	1563.00	6	2193.00	7	12152.00	6
Kerala	1487.67	9	1857.67	10	11475.33	7
MP*	1368.67	11	1617.33	12	7666.00	11
Maharashtra	2452.00	2	3573.00	2	15532.67	1
Orissa	1265.00	13	1463.00	14	6083.33	13
Puniah	2818.33	1	3828.67	I	15526.00	2
Raiasthan	1261.00	14	1890.67	9	8621.33	10
Tamil Nadu	1555.00	7	2290.00	5	12844.00	5
UP*	1299.33	12	1631.00	11	5757.33	14
West Bengal	1727.00	5	2235.67	6	10979.67	9
Av.	1669.16		2233.56		10488.91	
CV	31.56		36.79		37.44	

Note: Figures for 1981-82 to 1992-93 are at 1980-81 prices and remaining are at 1993-94 pnces.
* Undivided States.

Source: Based on data collected from Handbook of Statistics on Indian Economy, RBI 2005).

It is also observed that the better performing States in terms of NSDP are mostly in the western and southern regions of the country. On the other hand, poorly performing States are in the northern and eastern regions. Agriculturally most developed State—Punjab which occupied the first rank in PNSDP, in 1990-92, have relegated to the second rank in 2001-03, by Maharashtra. Relatively slower growth in the NSDP and higher population growth in the northern and eastern States are the main factors accountable for their lower PNSDP.

The Annual Compound Growth Rates in agriculture and non-agriculture sector of the States is presented in Table 2. A perusal of the table reveals the gloomy performance of most of the States during the last two decades which is more pronounced in the later decade. To the dismay, none of the States achieved growth rates more than four per cent and seven per cent in agriculture and non-agriculture sector respectively, during 1991-92 to 2003-04. The best performing States (in both agriculture and non-agriculture sector) during 1981-82 to 1990-91 namely, Haryana, Maharastra and Rajasthan are among the worst performers in agriculture sector during 1991-2003, though they retained higher growth rates in non-agriculture sector. Punjab, which experienced the highest growth rate in agriculture (4.91 per cent) during 1981- 1990 could achieve only a little more than two per cent during 1991-2003.

Table 2
Annual Compound Growth Rates of Agriculture and Non-Agriculture NSDP at Factor Cost (Constant Prices, 1993-94)

1981-82 to 1990-91			*1991-92 to 2003-04*	
Agri.	*Non-Agri.* < 7%	*Non-Agri.* > 7%	*Non-Agri.* < 7%	*Non-Agri.* > 7%
> 4 %	Punjab, WB	Haryana, Maharastra, Rajasthan	Bihar	–
3 %-4 %	Bihar, T N	–	–	W B
2%-3%	Assam, Kerala, M P	Karnataka, U P	AP, Punjab U P	Karnataka, Maharastra
< 2 %	–	A P, Orissa, Gujarat	Assam, M P, Orissa, T N	Gujarat, Kerala, Haryana, Rajasthan

Source: Based on data collected from Handbook of Statistics on Indian Economy, RBI (2005)

It is also evident from the table that all the States except Bihar and Andhra Pradesh have achieved much higher growth rates in agriculture sector during 1981-1990 than in 1991-2003. The only welcoming fact that can be drawn from the table is Bihar's remarkable agricultural growth during 1991-2003 which is above four per cent. The table shows that during the prereform period, almost 50 per cent of the total States achieved over 3 per cent annual growth in agriculture while during the post-reform period, except for Bihar and West Bengal, all the other States could achieve less than three per cent agricultural growth.

Table 3
NSDP at Factor Cost, Percentage of Workforce Engaged and Per Worker NSDP from Agriculture and Non-agriculture Activities (at 1993-94 Prices)

	NSDP in 2002-03 (Rs. crores)		*Percentage of Workforce (as per 2001 census)*		*Per Worker NSDP in 2002-03*		
State	*From Agri.*	*From Non-agri.*	*In agri.*	*In Non-agri.*	*From agri.*	*From Non-agri.*	*Ratio of Non-agri to agri.*
AP	21037	61010	62.16	37.84	9698	46212	4.77
Assam	6343	10445	52.36	47.64	12701	22986	1.81
Bihar*	20411	36696	74.45	25.55	7199	37708	5.24
Gujarat	11449	63997	51.58	48.42	10442	62186	5.96
Haryana	9440	22512	51.29	48.71	21969	55168	2.51
Karnataka	16114	47854	55.71	44.29	12291	45908	3.74
Kerala	6865	30171	22.80	77.20	29275	38005	1.30
MP*	14970	43646	72.85	27.15	5797	45312	7.82
Maharashtra	24207	129222	54.96	45.04	10698	69681	6.51
Orissa	6087	15775	64.77	35.23	6582	31364	4.77
Punjab	15022	22560	38.95	61.05	42258	40484	0.96
Rajasthan	10455	34314	65.91	34.09	6674	42348	6.35
Tamil Nadu	11482	69536	49.33	50.67	8349	49231	5.90
UP*	38430	64077	65.48	34.52	10276	32494	3.16
West Bengal	22875	66917	44.15	55.85	17574	40642	2.31
Average	14700	47916	55.12	44.88	14119	43982	3.12
COV	58.99	62.36	24.47	30.08	71.40	27.09	67.95

Source: Agriculture census various issues, and *www.indiastar.com*
* Undivided States

Inter-State comparison of agriculture and non-agriculture sectors is also made and the results are shown in Table 3. It is observed that NSDP from the agriculture and allied activities is much lesser than that from non-agricultural activities in all the States, whereas workforce dependence on agriculture is much higher than the non-agriculture sectors. Average figures for the 15 States show that about 55 per cent of the total workforce engaged in agriculture generates Rs.14700 crores of NSDP. Contrary to this, about 45 per cent manpower in nonagriculture sectors contributes NSDP of Rs.47916 crore which is 3.26 times of that of agriculture. This clearly indicates the poorer economic condition of farm *vis-à-vis* non-farm workforce.

Looking at the individual States, we find that workforce dependence on agriculture is high in economically backward States, such as Bihar (74.45%), MP (72.85%), Rajasthan (65.91%), UP (56.48%) and Orissa (64.77%). Kerala has the lowest proportion of workforce in agriculture (22.80%). It is followed by Punjab (38.95%), TN (49.33%) and WB (44.15%). The difference in NSDP from agriculture and non-agriculture activities is found largest in Orissa,followed by Gujarat and Maharashtra. The difference in Orissa is mainly due to very low agricultural income; while in Gujarat and Maharashtra, it is due to very high non-agricultural income.

We have also estimated per worker NSDP for agriculture and non-agriculture activities.The figures shown in Table-3 reveal a wide difference in per worker income from agriculture and non-agriculture activities. The difference is found substantial in MP, Maharashtra, Rajasthan, TN, Gujarat and Bihar. Per worker income from agriculture is lowest in MP, followed by Orissa and Rajasthan whereas non-agriculture income is worked out to be highest in Maharastra, followed by Gujarat, Haryana and TN. Punjab is the only state where per worker agriculture income is larger than that of non-agriculture. Kerala, Haryana and WB also have relatively higher per worker income from agriculture.

The table also evinces that disparity in per worker NSDP not only exists between agriculture and non-agricultural sectors but also among States. In case of agriculture, per worker NSDP is lowest in MP, followed by Rajasthan, Orissa and Bihar, whereas it is found highest in Punjab, distantly followed by Kerala, Haryana and West Bengal. Per worker NSDP from non-agricultural activities is observed highest in Maharashtra, Gujarat, Haryana, Tamil Nadu and Andhra Pradesh and lowest in Assam, Orissa, Uttar Pradesh and Bihar. Values of coefficient of variation show that disparity in per worker NSDP is much higher in agriculture than in the non-agricultural activities.

DISPARITY IN AGRICULTURAL OUTPUT AMONG STATES

In order to examine trend in the inter-state disparity in agriculture output, we have calculated Gini Coefficients for NSDP and PNSDP from agriculture for the last 23 years (1981- 82 to 2003-04) and results are shown in Figure 1. It is evident from the figure that though, variations in NSDP and PNSDP have declined over the years; disparity continues to be very high across the States. It is also observed that the variation in the NSDP is of higher magnitude than that of PNSDP. There is a strong evidence of cyclical nature of the disparity across States.

Figure 1: Gini Coefficient of NSDP and PNSDP in Agriculture from 1981-2003

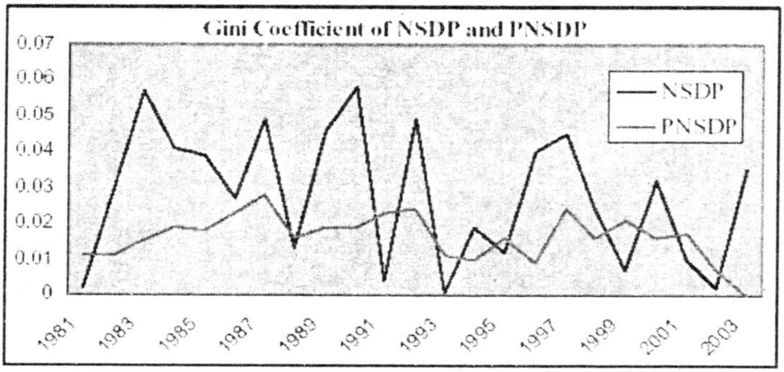

It may seen from Table 4 that seven States namely, Haryana,, Maharashtra, Punjab,Tamil Nadu, Rajasthan, West Bengal and Bihar had recorded a higher growth rate in agricultural NSDP than the all India average growth rate of

Table 4
Annual Compound Growth Rates of NSDP (Agri.) and Population Growth

	NSDP from Agriculture		*Population Growth #*	
States	*1981-82 to*	*1991-92*	*1981-91*	*1991-2001*
A.P.	1.71**	2.79*	2.19	1.72
Assam	2.19*	1.03*	2.14	2.13
Gujarat	-0.97	0.69	2.02	1.76
Haryana	4.90*	1.78*	2.46	2.19
Karnataka	2.04*	2.45*	2.03	1.49
Kerala	2.72*	-1.06*	1.42	1.35
M.P.⊕	2.16*	1.17	2.38	1.98
Maharashtra	4.23*	2.71*	2.29	1.97
Orissa	1.26	0.39	1.82	1.81
Punjab	4.91*	2.81*	1.88	1.91
Rajasthan	4.33**	1.49	2.64	2.06
Tamil Nadu	3.32*	0.29	1.48	0.98
U.P.⊕	2.71*	2.54*	2.33	1.75
West Bengal	6.04*	3.86*	2.20	1.72
India	3.05*	1.48*	2.16	1.88

⊕ Undivided States.

Source: Based on data collected from Handbook of Statistics on Indian Economy, RBI (2005)
Bhattachrya and Sakthivel (2002)
* and ** indicate 1 per cent and 5 per cent level of significant, repectively.

3.05 per cent during the 1980s. Gujarat had negative but insignificant growth whereas Andhra Pradesh and Orissa achieved less than two per cent growth. Rajasthan, West Bengal, Haryana, Maharashtra and Punjab have progressed rapidly during the 1980s with over four per cent per annum growth. However, during the post reform period, agricultural NSDP grew over three per cent per annum only in Bihar and WB. Punjab, AP, Maharashtra and UP are the other major States which have performed well by achieving a growth rate of over two per cent per annum. Rest of the states either had negative or positive but insignificant agricultural growth in 1990s. A comparison of agricultural growth with the population growth shows that in the prereform decade, except for four states, namely AP, Gujarat, MP and Orissa, all the other states achieved agricultural growth rate higher than the population growth rate whereas in the decade of economic reform, population growth in eight states, viz., Assam, Gujarat, Haryana, Kerala, MP Orissa, Rajasthan and TN grew faster than the agricultural output (NSDP). At national level also, population growth between 1991 and 2001 was higher than the corresponding growth in the agricultural output.

REASONS FOR DISPARITY

A number of factors may be accountable for disparity in the agriculture output across States. Agriculture development depends on various factors, such as size of operational holdings, irrigation facilities, cropping intensity, application of new farm technology, HYV seeds, fertilizers, pesticides, extension services, and rural infrastructure, including road, energy, credit and market. Besides these factors, quality of land, sunshine and rainfall can also explain the difference in agricultural productivity. Above all, it is the capacity and willingness of individual State governments to invest in agriculture that determines the variation in the agricultural growth, as agriculture is a state subject. State policy related to the agricultural development, therefore, plays a vital role in explaining variation in agricultural productivity across state. In this section, we have identified some indicators of agricultural disparity across states.

Disparity in State Governments' Expenditure on Agriculture

State-wise plan and non-plan revenue expenditure on agriculture and allied activities as percentage of total plan and non-plan revenue expenditure is shown in Table-5. It is observed that the share of agriculture in the total revenue expenditure (both plan and non-plan) has significantly declined during the last one and a half decade in all the States under reference. On an average, share of agriculture in the total plan expenditure on revenue had declined from 15.66 per cent in 1991-92 to 10.81 per cent in 2005-06, whereas the agriculture's share in the non-plan revenue expenditure decelerated from 6.38 per cent to 3.44 per cent between the same periods. If we

Table 7
Changes in Percentage Distribution of Number of Operational Holdings

	1991-92					2002-03				
State	*M*	*S*	*ME*	*L*	*Av. Size*	*M*	*S*	*ME*	*L*	*Av. Size*
AP	59.3	21.4	18.6	0.8	1.29	60.7	20.7	17.5	1.1	1.33
	(17.5)	(23.3)	(49.7)	(9.4)		(18.6)	(21.1)	(44.9)	(15.5)	
Assam	70.8	20.0	9.0	0.2	0.88	76.2	18.4	5.3	0	0.69
	(34.2)	(31.2)	(32.0)	(2.6)		(42.0)	(36.0)	(22.0)	(0.0)	
Bihar*	76.8	13.7	9.4	0.2	0.75	82.6	12.2	5.0	0.2	0.59
Guj.	47.9	19.9	29.8	2.5	1.99	60	17.3	20.9	0.8	1.58
	(8.5)	(13.8)	(59.9)	(17.8)		(13.1)	(15.0)	(56.3)	(15.6)	
Hary.	50.7	13.5	31.8	4.0	2.19	66.3	12.8	20.1	0.9	1.28
	(5.3)	(8.8)	(54.9)	(31.0)		(10.4)	(13.5)	(61.0)	(15.1)	
Karn	49.7	20.3	27.8	2.3	1.58	58.2	20.4	20.3	1.1	1.38
	(9.6)	(15.4)	(56.0)	(19.0)		(16.2)	(20.0)	(52.6)	(11.1)	
Kerala	91.6	5.9	2.5	0	0.35	91.8	6.2	2.0	0	0.37
	(53.3)	(23.4)	(56.0)	(23.0)	(0.04)	(57.8)	(23.3)	(18.9)	(0.0)	
MP*	38.7	24.4	34.4	2.5	2.24	51.2	23.3	24.4	1.2	1.58
	(6.7)	(15.6)	(61.4)	(16.4)		(13.1)	(20.3)	(27.0)	(11.2)	
Mah	43.6	18.9	34.5	3.0	2.25	49.5	21.4	27.9	1.2	1.65
	(6.7)	(11.8)	(61.3)	(20.3)		(12.0)	(17.7)	(59.6)	(10.7)	
Orissa	60	24.3	15.4	0.3	1.13	78.4	15.2	6.3	0.1	0.72
	(22.1)	(30.2)	(44.1)	(3.7)		(43.0)	(28.7)	(27.4)	(0.9)	
Punjab	63.2	11.4	23.7	1.7	1.46	66.3	11.2	20.7	1.9	1.31
	(6.2)	(10.7)	67.3)	(15.8)		(7.3)	(11.7)	(62.6)	(18.5)	

Contd...

State	*M* (1991-92)	*S* (1991-92)	*ME* (1991-92)	*L* (1991-92)	*Av. Size* (1991-92)	*M* (2002-03)	*S* (2002-03)	*ME* (2002-03)	*L* (2002-03)	*Av. Size* (2002-03)
Raj	39.3	19.9	33.7	7.1	3.08	49.4	18.5	27.4	4.7	2.36
	(5.6)	(9.4)	(47.5)	(37.7)		(9.0)	(10.9)	(47.0)	(33.1)	
T N	77.2	14.1	8.4	0.3	0.71	77.1	13.4	9.4	0.1	0.76
	(28.9)	(28.1)	(37.9)	(5.1)		(30.9)	(24.2)	(43.4)	(1.5)	
UP*	68	18.5	13.2	0.3	1.01	76.7	15.9	7.3	0.1	0.75
	(25.0)	(26.3)	(44.5)	(4.3)		(35.7)	(29.2)	(32.3)	(2.8)	
W.B	80.7	13.4	5.9	0	0.60	88.8	8.9	2.3	0	0.44
	(40.0)	(30.7)	(29.4)	(0.0)		(58.3)	(26.7)	(14.9)	(0.0)	
Avg.	61.2	17.3	19.9	1.7	1.43	68.9	15.7	14.5	1.0	1.12
	(19.9)	(20.3)	(47.4)	(12.5)		(27.4)	(21.7)	(41.7)	(9.3)	
COV.	26.5	29.6	58.0	117.7	53.98	20.3	31.3	64.7	128.5	49.92
	(74.9)	(41.0)	(27.5)	(91.8)		(65.8)	(33.2)	(39.8)	(101.5)	

Source: Agriculture census various issues
Figures in parentheses are percentage distribution of area operated by respective size groups.

The value of COV in the average size shows that though the disparity across the state was high, it marginally declined in 2002-03 over 1991-92.

Size of holding is one of the vital factors in explaining variation in the agricultural output. Marginal and small farmers face problem in having access to modern farm inputs and also suffer from diseconomies of small size. Here the purpose is not to debate on whether the productivity of small farm is higher than the large farms; the aim is to demonstrate that modern farm practices can be more cost-effective if the size of land holding is relatively large. New farm technology may be size-neutral, as has been the case of green revolution technology, but it is not certainly resource-neutral. This has been well documented in the literature that large farmers have better access to technology, inputs, credit, market and subsidies.

Disparity in Input Intensification

Table 8 provides State-wise information about the level of use of three most important modern inputs in agriculture namely, fertilizer, tractor and irrigation during 1981 to 2003. The extent of disparity across States in using these modern inputs is brought out by the coefficient of variation during three sub-periods, Triennium Ending (TE) 1981-83, 1991-93 and 2001-03. Wide disparity across States in the level of use of all the three inputs is observed during all the sub-periods. Though it is decreasing over period the magnitude continues to be very high in all the cases. In the case of fertiliser consumption, the COV was 77.83 per cent during 1981-83 decreased to 69.13 per cent in 1991-93. It further declined perceptibly to 49.72 per cent in 2001-03. The disparity across States in tractor use found to be highest among these three inputs to the extent of 125.22 per cent, 151.78 per cent and 97.23 per cent during the three sub-periods respectively. Wide disparity is also seen in the irrigated area. Since, fertiliser and irrigation are complementary inputs they show a somewhat similar trend.

Inter-State disparity in access to and utilization of farm inputs is also noticed. Energy is one of the vital inputs that affect the agricultural output significantly. Table 9 shows state-

Table 8
State-wise Level of Input used during 1981-2003

	Fertliser Consumption (Kgs./ha)			*No. of Tractor Use/000 ha*			*Percentage of GCA Irrigated*		
States	*1981-83*	*1991.93*	*2000.03*	*1981.83*	*1991.93*	*2000-03*	*1981-83*	*1991-93*	*2000-03*
AP	60.36	147.14	153.23	1.52	5.29	6.81	33.58	43.11	42.26
Assam	4.76	13.12	38.65	0.19	0.26	0.37	11.59	15.56	5.49
Bihar	22.83	103.89	87.98	1.53	1.88	9.11	34.74	44.00	48.25
Guj	46.65	77.70	85.26	2.81	7.01	13.75	21.95	29.13	32.92
Hary	61.67	182.51	155.75	7.89	44.74	30.66	63.40	83.71	75.46
Kar	45.48	84.26	107.06	2.18	3.69	5.25	16.23	23.24	25.89
Kerala	48.02	99.81	59.20	0.58	0.89	0.71	10.98	12.96	14.68
M P@	13.61	44.27	42.18	1.19	2.58	4.64	13.68	3.90	26.20
Mah	31.30	70.33	75.47	1.10	4.63	10.43	12.23	15.12	17.68
Orissa	14.73	31.01	39.17	0.27	0.36	2.09	28.96	29.34	26.03
Punjab	141.76	302.11	176.25	14.19	50.99	38.18	92.24	94.77	96.77
Raj	7.77	27.19	37.94	3.35	8.99	23.65	21.96	28.11	34.09
TN	92.58	151.30	149.46	2.28	5.32	9.60	48.65	48.71	53.64
UP*	56.29	130.45	128.36	5.6	20.10	26.18	49.37	63.05	68.02
W B	40.68	141.63	120.48	0.27	1.22	1.94	24.55	25.56	42.25
Av.	45.90	107.12	97.10	2.99	10.53	12.22	32.35	37.35	40.64
COV	77.83	69.13	49.72	125.22	151.78	97.23	69.93	69.92	60.71

Source: Agriculture census various issues; and *http://icar.naarm.ernet.in/*
* Undivided states

wise percentage of farmers using diesel, electricity and animal energy in ploughing, harvesting and irrigation operations. The figures presented in the table show that the proportion of farm households using diesel operated implements (tractor operated machines) in ploughing was relatively high in the States of Punjab, Haryana, Uttar Pradesh, Kerala and Rajasthan, while it was very low in Jharkhand, Assam, Chhttisgarh, Orissa and Karnataka. In case of harvesting of crops by diesel operated machines, the percentage of farm households was found highest in Haryana (99%), followed by Punjab (96%) and Uttar Pradesh (87%) and lowest in Jharkhand (0%), followed by Assam (1%) and Orissa (7%).

Table 9

Inter-state Variation in the Proportion of Households Reporting Use of Animal, Diesel and Electricity in Various Farm Operations (2003)

	% of Farmer HHs ploughing land using		*% of Farmer HHs Irrigating land using*		*% of Farmer HHs haresting Crops using*	
State	*Diesel-operated Implements*	*Animal-operated Implements*	*Diesel Pumps*	*Electric Pumps*	*Diesel operated Implements*	*Animal operated Implements*
AP	38	61	20	78	60	40
Assam	3	97	87	4	1	99
Bihar	64	35	97	2	51	43
Chhattisgarh	10	90	28	63	10	90
Gujarat	52	47	35	63	59	35
Haryana	89	11	53	47	99	1
Jharkhand	2	98	81	2	0	100
Karnataka	14	85	7	89	27	72
Kerala	72	27	15	85	61	8
MP	34	65	34	65	79	16
Maharashtra	21	78	12	87	44	46
Orissa	12	88	61	38	7	93
Punjab	96	3	29	71	96	0
Rajasthan	70	29	61	34	79	17
Tamil Nadu	68	30	27	72	79	13
U.P.	81	18	84	16	87	8
West Bengal	53	47	87	13	23	74
India	47	52	66	33	59	38
COV	68.60	59.28	62.79	64.89	66.78	82.18

Source: Access to Modern Technology for Farming, Situation Assessment Survey of Farmers, 59th Round, Report No. 499(59/33/2),

It is also observed that the percentage of farmers using electric pumps in irrigating crops was highest in Karnataka (89%), closely followed by Maharashtra (87%) and Kerala (85%). Andhra Pradesh, Tamil Nadu and Punjab also had relatively high percentage of farm households using electric pumps in irrigating crops. On the contrary, this percentage was very low in Bihar, Jharkhand, Assam, West Bengal and Uttar Pradesh. A wide disparity in percentage of farm households using diesel and electricity in various farm operations implies that the benefits of subsidies on diesel and power are limited to the farmers of those regions which use of these energy sources is relatively high.

DETERMINANTS OF DISPARITY

Based on the priory logic, data accessibility and review of relevant literature on agricultural disparity, we identified four explanatory variables for studying the variation in the agricultural NSDP. However, after conducting a partial correlation test, two variables, namely, cropping intensity (CI) and number of tractor use (TRAC) were found to be highly correlated with each other posing the problem of multicollinearity. Therefore, we had to run two regression models separately, where in model-I we have included variation in fertilizer application, cropping intensity and rural infrastructure as the explanatory variables. In model–II variation in fertilizer application, farm mechanization and rural infrastructure are included as explanatory variables. The variables selected for the regression models are:

FERT : Coefficient of Variation (COV) in fertilizer use across States

LPR : COV in length of pucca road (proxy for rural infrastructure)

TRAC : COV in number of tractor use (proxy for mechanization)

CI : COV in cropping intensity.

The Ordinary Least Square (OLS) model of the following functional form has been applied.

$$NSDP_t = \alpha + \sum_{j-1}^{n} \beta_j X_{jt} + \mu_t$$

Where NSDP stands for coefficient of variation in agricultural NSDP; j for exogenous variables; and t for year of observations (1981-82 to 2002-03); α and β_j are parameters to be estimated; μt is the random unobserved disturbance with zero mean and a constant variance.

It is evident from table-10 the regression results that both the equations are statistically significant in terms of goodness of fit. Values of R^2 are fairly high, to the extent of 48 and 69 per cent. F-values for both the equations are also found significant indicating the usefulness of the model. The results of model- I reveal that 69 per cent variation in the COV of NSDP is explained by the variables included in the equation. Turning to the individual variables, we observe that, the variables FERT and LPR are statistically significant and have positive impact on the COV of NSDP. The variable CI also has a positive impact on the dependent variable though not statistically significant. On an average, a one per cent increase in the variation of fertilizer use enhances the variation in NSDP by 0.21 per cent.

Table 10
Estimates of Regression Analysis

Variables	*Constant*	*FERT*	*LPR*	*CI:*	*TRAC*	*F-Value*	R^{-2}
Equation I	0.922	0.023*	0.508*	0.241	–	16.74*	0.69
	(0.712)	(0.006)	(0.086)	(0.179)			
Equation II	0.831	0.021*	0.434*	–	0.233	15.97*	0.68
	(0.936)	(0.006)	(0.090)		(0.218)		

Notes: (1). * indicates 1 per cent level of significant.
(2) Figures in parentheses are standard errors of the respective coefficients.

Another important determinant of variation in agricultural income is the variation in the length of pucca road. A better road communication enhances the accessibility of farmers to the markets of urban areas which consequently fetches higher price of their produces and thereby increases the agricultural income. Therefore, a variation in the length of pucca road is expected to have a positive impact on the variation in income. A one per cent increase in CV in LPR enhances the CV in NSDP by 0.51 per cent.

Equation-II also produces a similar kind of result showing the positive impact of variation in the number of tractor use on the CV of NSDP. Magnitude of its coefficient shows that one per cent increase in the CV of number of tractor use increases the CV of NSDP by 0.23 per cent. As tractor use is taken as a proxy variable for farm mechanization and farm mechanization has significant impact on raising agricultural production and productivity and reducing the cost of cultivation and consequently raising the agricultural income, positive impact of variation in the tractor use on the variation in the income is quite a meaningful finding.

CONCLUSIONS

It is warranted to identify the reasons for the variations in the agricultural income across the States for achieving equitable growth of income and employment in the rural farm sector. Therefore, this paper examines the extent of disparity that exists in the agricultural sector across the fifteen major States of India; identifies the major factors in explaining this disparity; and finally suggests policy interventions to narrow down the inequality among regions. On the basis of data collected from various secondary sources for the period 1981-2003, the paper presents strong evidence of inequality in agricultural income across States of India. It is also observed that the better performing States are mostly in the western and southern regions of the country. On the other hand, the non-performing or poorly performing States are in the northern and eastern regions. The study also focuses on the growing disparity in the public investment in agriculture.

The study finds that though there is a decrease in the disparity across the States in both NSDP and PNSDP over the years, the disparity continues to be very high and it is relatively higher in the NSDP as compared PNSDP; probably due the sharp reduction in the population growth rate in the economically poor States. There is a strong evidence of cyclical nature of the disparity across the States, implying that agriculture still is highly dependent of the nature and other stochastic factors that cause fluctuations in the agricultural output across time and space. It is also observed that agricultural growth has decelerated in the agriculturally developed States like Punjab, Haryana and Western Uttar Pradesh in the post-reform period and some agriculturally poor states such as Bihar and West Bengal have shown a significant improvement in their agricultural growth. As growth rate is pattering off in the green revolution regions and picking up in the other States, especially in eastern and north-eastern regions, there is likelihood that in the years to come inter-state disparity would be narrowed down provided that the state governments make sufficient investment in irrigation & flood control, power, road, agricultural R&D and extension services.

Empirical results show that disparity in the agricultural output significantly and positively affected by the variations in fertilizer use and in length of pucca road. The results also confirm that variation in the number of tractor use and in the cropping intensity have positive impact on the variation in agricultural output; however, the coefficients are not found statistically significant.

POLICY IMPLICATIONS

The following policy implications emanate from the findings of the study:

(1) Agricultural disparity is found positively associated with the variation in the level of infrastructure (pucca road), fertilizer consumption and farm mechanisation. It is therefore necessary to improve the infrastructure in the infrastructure-deficit state and provide easy access to the farmers for short term institutional credit facilities to the farmers to purchase fertilizer, seeds etc. and long term credit to procure farm machines such as tractor. As there is significant difference in the usage of modern farm inputs among states, efforts should be made to remove both the supply side and the demand side constraints in the access to these inputs.

(2) The cyclical pattern of agricultural disparity across states indicates that natural and other stochastic

factors like flood and drought have significant impact on the agricultural output. The state governments have to make substantial enhancement in watershed management, irrigation and flood control, especially in those state where frequency of such occurrences is very high.

(3) About 85 per cent of operational holdings in India are below 2 hectare, with the percentage as high as 98 in Kerala, 97 in West Bengal, 95 in Bihar and 94 in Orissa. There is a need to make these holdings economically viable so that their productivity and profitability may be increased. Promotion of "group farming" or " collective farming" among small and marginal farms through active support of government institutions, commercial banks and non-governmental organizations may go a long way in reducing the scale-disadvantages of these farms and in enhancing the farm productivity and profitability and reducing disparity not only across size of farms but also across regions.

REFERENCES

Bhanumurthy N. R. and Sapna S. (2004), 'Industrial Recovery: Can It Be Sustained', *Economic and Political Weekly,* Vol. 39, No. 5.

Bhattacharya, B. B. and S. Sakthivel (2004), 'Regional Growth and Disparity in India', *Economic and Political Weekly,* (EPW), March 6, pp 1071-1077.

Chand, Ramesh. (2004). Agricultural Growth during the Reforms and Liberalization: Issues and Concerns. Policy Brief 20. New Delhi, National Centre for Agricultural Economics and Policy Research.

Das, T. K. (2002), 'Convergence across Indian States: Cross-Sectional and Panel Estimations', *South Asia Economic Journal,* Volume 3, No. 2, pp. 227-239.

Joshi, P. K., Birthal P. S. and Minto N. (2006), 'Source of Agricultural Growth in India: Role of Diversification towards High Value Crops', MTID Discussion Paper No. 98, International Food Policy Research Institute, Washington, D.C. 20006 U.S.A

Kurian, N. J. (2000), 'Widening Regional Disparities in India: Some Indicators', *Economic and Political Weekly,* (EPW), Vol. 35, No 7, pp. 538-55.

NSSO. (2005), Some Aspects of Farming, Situation Assessment Survey of Farmers, 59th Round, Report No. 496 (59/33/3), Ministry of Statistics and Programme Implementation, Government of India, New Delhi.

NSSO (2005), 'Access to Modern Technology for Farming, Situation Assessment Survey of Farmers', *59th Round, Report No. 499(59/33/2),* Ministry of Statistics and Programme Implementation, Government of India, New Delhi.

Shetty, S. L. (2003), 'Growth of SDP and Structural Changes in State Economies:Inter-StateComparisons', *Economic and Political Weekly,* (EPW), Vol. 38, No. 49, December 6, pp. 5189-5200

Singh, Nirvikar, L. Bhanadari, A. Chen and a. Khare, (2003), "Regional Inequality in India: A Fresh Look", *Economic and Political Weekly,* (*EPW,* Vol. 38, No. 11, March 15, pp. 1069-1073.

Singh, S. P. (2005), 'Agriculture', in Disequalising Growth: Alternative Economic Survey, India, New Delhi, Daanish Books.

Singh S. P. (2006), "Agriculture" in Disequalising Growth: Alternative Economic Survey, India, Danish Rainbow Publishers, New Delhi, pp. 53-68.

6

DEVELOPMENT OF AGRO-BASED INDUSTRIES IN ORISSA: WAY TO HIGHER AGRICULTURAL PRODUCTIVITY

MINATI MALLICK

INTRODUCTION

The importance of agriculture can hardly be undermined in a state like Orissa where this sector contributes 23.43 per cent of State Domestic Product and provides employment to 65 per cent of total work force. The sector so to say is the mainstay of the poverty population of the state. The dream of improving the economic status of the rural masses will be a myth unless and until the growth of its agriculture is enhanced from the present dismal rate of less than 2 per cent.

Increasing agricultural labour force, lower capital formation, very small per capita availability of cultivable land (0.15 ha), predominant presence of marginal and small farmers, inadequate infrastructural facilities, marketing and credit problems, lack of extension services are main impediments to agricultural growth. Corrective measures are highly essential for achieving 4 per cent growth rate as targeted in current plan to support the increasing population of the state.

Economic progress no doubt depends highly on the pace of industrialization and deployment of surplus rural labour in industrial and services sector but the capital intensity of industrialization poses a great challenge to an agrarian state like Orissa. Moreover, rapid rural urban migration leads to growth of an informal sector and creates difficulties in provision of basic amenities of life in that sector as it requires huge amount of investment.

The solution therefore lies in establishment of agro-based industries that help withdrawing surplus labour engaged in agriculture and reducing man-land ratio thereby increasing agricultural productivity. The capacity of this sector in generating employment with a short gestation period coupled with less capital intensity has made it an ideal strategy for economic development of the backward agrarian states like Orissa.

Moreover, the agro-based industries facilitate the use of agricultural products as raw materials providing outlet to surplus produce in the sector. By widening the market, it induces the farmers to raise production and productivity. In addition to this, the high export potential of the processed agri-products also ensures widened market and alluring remunerations for the farmers engaged in the production of exportable. The establishment of agro-based industries inculcates industrial culture, develops entrepreneurship in rural areas. It also generates a sense of participation in the development process.

The present paper that attempts to study the importance of agro-based industries in promotion of agricultural productivity is organized in the following way. Section-II contains the objectives of the study. The materials and methods are presented in Section-III. Results of the study are discussed in Section-IV followed by the concluding Section.

OBJECTIVES

The broad objectives of the present study are as follows:

(1) To find out the growth trend of agro-based and non-agro based industries in the state.

(2) To find out the elasticity of employment with respect to investment in both agro-based and non-agro based industries.

(3) To suggest measures for balanced growth of agro-based industries in the state.

MATERIALS AND METHODS

The present study makes use of secondary data collected from the annual issues of Statistical Hand Book published by the Directorate of Industries, Orissa, Cuttack. The data relate to the number of SSI units, investment and employment. The study covers the period from 1984-85 to 2000-01 for which comparable data are available. It could not be extended further because of non-availability of data and it is assumed that the trend remains the same.

In order to find out the growth trend of number of SSI units, investment and employment, an exponential trend line is estimated by using the following formula.

$$Y = Ae^{bt}$$

$\Rightarrow$ $(dY/dt)/Y = b$,

where b = exponential growth rate

The investment elasticity of employment is obtained by estimating a log linear function.

$$\ln E = \alpha + \beta \ln I,$$

where E = employment

I = investment

α, β = parameters

Therefore, $d(\ln E)/dI = d/dI\,(\alpha + \beta \ln I)$

$\Rightarrow$ $(1/E) \times (dE/dI) = \beta/I$

$\Rightarrow$ $\beta = (dE/dI) \times (I/E)$

β = investment elasticity of employment

RESULTS AND DISCUSSION

Number of SSI Units

The number of SSI units took a gigantic leap forward during the period under reference. It is evident from the Table 1 that the total number of SSI units, which stood at 23,437 in the year 1984-85, increased to 66,206 in the year 2000-01. The number of agro-based industries is found to have increased from 5777 to 15,551 while that of other industries increased from 17,660 to 50,655 during the same period.

Table 1
Number of SSI Units

Years	*Agro-based*	*Others*	*Total*
1984-85	5777	17660	23437
1985-86	6447	20472	26919
1986-87	6978	22454	29432
1987-88	7479	24245	31724
1988-89	7893	25967	33820
1989-90	8330	27515	35845
1990-91	8805	29289	38094
1991-92	9270	31057	40327
1992-93	9762	32682	42444
1993-94	10285	34470	44755
1994-95	10811	36271	47082
1995-96	11319	38270	49589
1996-97	11938	40749	52687
1997-98	12689	43184	55873
1998-99	13558	45499	59057
1999-2000	14599	47931	62530
2000-01	15551	50655	66206

Source: Statistical Hand Book.

Table 2
Exponential Growth Rate of Number of Industries
(Regression Results)

Parameters	*Agro based*	*Non-Agro based*
A	5794.91	18509.10
b	0.0573	0.0609
R^2	0.994	0.987

The exponential growth rate of the number of agro-based industries is calculated to be 5.73 per cent which is lower than 6.09 per cent for that of other industries as shown in Table 2.

Investment

Investment plays a vital role in determining the level of employment and output. More the level of investment more will be the output and employment of resources both physical and human. The speed and pattern of growth of industries are also closely related to the volume of investment.

Table 3
Level of Investment in SSIs (Rs. in Lakh)

Year	*Agro based*	*Others*	*Total*
1984-85	5229.46	15018.19	20247.65
1985-86	5982.92	18904.83	24887.75
1986-87	6852.40	22542.06	29394.46
1987-88	7550.97	26110.38	33661.35
1988-89	8570.61	30000.02	38570.63
1989-90	9809.85	33787.05	43596.90
1990-91	11033.35	38623.42	49656.77
1991-92	11988.77	42871.38	54860.15
1992-93	13198.25	47161.86	60360.11
1993-94	14349.14	51631.61	65980.75
1994-95	16838.38	55950.52	72788.90
1995-96	19628.71	60642.09	80270.80
1996-97	23131.07	67592.25	90273.32
1997-98	27570.34	76561.68	104132.02
1998-99	31133.27	92005.01	123138.28
1999-2000	35890.00	103542.10	139432.10
2000-01	41322.58	113427.22	154749.80

Source: Statistical Hand Book.

Table 3 reveals that there has been a phenomenal growth in investment in small scale industries.The level of investment has increased from Rs. 20, 247.65 lakh to Rs. 1, 54,749.80 lakh during the study period. It may be noted here that the investment in agro-based industries has increased from Rs. 5,229.48 lakh to Rs. 41,322.58 lakh registering an exponential growth of 12.67 per cent. On the other hand, the investment in other industries has increased from Rs. 15,018.19 lakh to Rs. 1, 13,427.20 lakh registering an exponential growth of 11.72 per cent.

Table 4
Exponential Growth of Investment
(Regression Result)

Parameters	*Agro based*	*Non-Agro based*
A	4496.59	15685.10
b	0.1267	0.1172
R^2	0.994	0.989

Employment

Employment in the SSI sector, as shown in the Table 5, has increased from 1, 82,091 in the year 1984-85 to 4, 35,598 in the year 2000-01. The corresponding figures in Agro-based vs. other industries stand in the order of 32,599 & 79,877 and 1, 49,492 & 3, 55,729. The exponential growth of employment in the agro-based industries is calculated to be 5.20 per cent as compared to 5.02 per cent for other industries for the period under reference.

Table 5
Level of Employment in SSIs

Year	*Agro based*	*Others*	*Total*
1984-85	32599	149492	182091
1985-86	35553	169962	205515
1986-87	38242	184250	222492
1987-88	40679	196641	237320
1988-89	42987	208250	251237
1989-90	45053	220233	265286
1990-91	47168	233775	280943
1991-92	49042	247446	296488
1992-93	51012	258820	309832
1993-94	53335	270304	323639
1994-95	56148	280587	336735
1995-96	58939	290815	349754
1996-97	62248	303135	365383
1997-98	66041	316058	382099
1998-99	70209	328666	398875
1999-2000	75255	342228	417483
2000-01	79877	355721	435598

Source: Statistical Hand book

Table 6
Exponential Growth of Employment
(Regression Results)

Parameters	*Agro based*	*Non-Agro based*
A	32260.20	15836.80
b	0.0520	0.0502
R2	0.994	0.977

Investment Elasticity of Employment

In view of the ever-increasing magnitude of unemployment and underemployment, it is imperative that the situation be controlled by undertaking employment- cum – production-oriented activities. Since the pressure on land is already high, it is now emphasized by all that the solution to this problem lies in rapid growth of small-scale industries in general and agro-based industries in particular. It is found that the investment elasticity of employment (0.409) is higher in agro-based industries compared to that (0.329) of other small-scale industries taken together. Thus, the employment generating capacity per unit of investment is more in agro-based industries compared to that of other industries as a whole.

Table 7
Investment Elasticity of Employment
(Regression Results)

	Constants	*Se*	*t*	*R^2*
Agro-based	α =1.310	0.015	86.62	0.994
Industries	β = 0.409	0.008	48.81	
Other	α =1.477	0.021	71.79	0.992
industries	β = 0.329	0.010	32.19	

FINDINGS AND SUGGESTIONS

The major findings that emerged from the study are as follows:

- So far as the number o f SSIs is concerned, agro-based as well as other industries are increasing with an exponential growth rate of 5.73 and 6.09 per cent respectively.
- In terms of investment, the agro-based industries are ahead of other industries with an exponential growth rate of 12.67 per cent during the study period.

- The employment in agro-based industries is also registering a higher growth rate of 5.20 per cent as compared to the other small-scale industries.
- With a higher investment elasticity of employment i. e.0.409, the agro-based industries emerge to be a potential sector of employment generation that can reduce man-land ratio thereby increasing agricultural productivity.

The recent industrial policy resolution of Orissa (2007) has put a robust policy framework to create an environment for industrial promotion and investment facilitation in the state. But agro-based industries find no special attention or there any clear-cut defined strategy for promotion of these strategies. In the context of present scenario in agriculture and the importance of agro-based industries in the state, the study makes the following suggestions for the development of these industries

- Promotion of those agro-based industries that make use of agri-products in which the state has competitive advantage should be encouraged.
- With strong feedback in terms of higher demand for raw materials, the agro-based industries will induce generation of surplus agri- products. Necessary steps should be taken to see that there is sufficient surplus of agricultural output in the state. This can be achieved through encouraging diversification of agriculture.
- A coordinated establishment of related groups of agro-based industries will ensure fruitful use of byproducts.
- Investment should be encouraged in all the districts with special emphasis on the less developed ones to ensure balanced and homogeneous growth of the agro- based industries in the state.
- The labour intensive agro- based industries should be encouraged in the districts with low density of population so that migration to high-density regions can be prevented.

- Extension services and dissemination of information along with research activities and training programmes are necessary. Steps should be taken in promotion of all these activities.

Since large scale industries does not automatically lead to an improvement in the level of living and over crowded agriculture does not hold any promise for rapid economic development, prosperity of the state lies in a balanced development of small scale industries which need to be linked with agriculture on the one hand and large scale industries on the other. The particular emphasis that we want to put here is that Agro-based industries that not only make use of local resources but also help generating employment and output catering to local demand and demand from outside. These industries have high degree of complementarities with agriculture sector. By withdrawing surplus labour on the one hand and creating market for agricultural products on the other, it helps increase in agricultural productivity and paves way for development of agriculture and the state.

REFERENCES

Lenka, J. and Mohanty, A. (2002), "An Economic Enquiry into the Determinants of Agro Industry Development in Orissa", in Mishra, B. (ed), *Agro Industries and Economic Development*, Deep & Deep Publications, New Delhi, pp. 32-41.

Mishra, B. (2002), "Role of Small Industries in Rural Development", in Mishra, B. (ed), *Agro Industries and Economic Development*, Deep & Deep Publications, New Delhi, pp. 3-17.

Mishra, S. N. (1999), "Can Agro Industries be an Effective Tool for Promoting Rural Development in Orissa?", *Journal of Rural Development*, Vol. 18, No. 2, NIRD, Hyderabad.

Statistical Hand Book, Directorate of Industries, Orissa, Various Issues.

7

FACTORS BEHIND THE EXTENT OF CROP DIVERSIFICATION IN THE PLAINS OF ASSAM: A SPATIAL ANALYSIS

RAJU MANDAL

INTRODUCTION

One of the major challenges in the 11th Five Year Plan is to reverse the deceleration in agricultural growth. The two major sources of growth in agriculture, viz., area expansion and productivity growth, which served well in the past, are now plagued with some limitations. It is argued that any significant technological breakthrough cannot be expected in the near future and hence, one has to depend on the exploitation of the potential of the existing technology (GOI, 2006). Therefore, a third alternative which may prove to be very useful in this context, at least in the short run, is to move towards diversification into high value crops.

The dominance of wheat-paddy system has led to serious economic, social and ecological problems such as deceleration in productivity growth, drop in agricultural employment, overexploitation of ground water resources and decline in soil fertility especially in the Green Revolution areas of the country. In Punjab soil has deteriorated because of excessive use of chemical fertilizers and growing the same crops over and over again (Sidhu and Dhillon, 1997). The over dependence of Punjab on rice and wheat has led to some serious ecological and production related marketing problems (Sidhu and Sidhu, 1988). Moreover, the gradual changes in the consumption pattern within the country as well as abroad are a clear indication that the policy focus needs to be reoriented towards diversifying the cropping pattern to meet the emerging market needs. Diversification of agriculture in favour of more competitive and high value quality products assumes greater importance in the context of opening up of the economy and its competitiveness, particularly in the farm sector.

The lesson for Assam from the experience of the Green Revolution areas in India is not to concentrate only on one or two crops but to go for a diversified cropping pattern to avoid the adverse environmental consequences apart from reaping other benefits such as stability of yields, reduced risk of crop failure, increased productivity, increased agro-returns, enhanced employment and so on. There is a special need of diversification of crops in Assam. The 'Sali' paddy that has traditionally occupied a major share in the total cropped area is very much prone to the frequent floods, which is a major source of instability in agricultural production in the state. Hence, diversification into not only crops but also seasons may provide a boost to agricultural production and its stability to a great extent[1].

Following deployment of tubewells in the 1990s there was a spurt in the production of summer (boro) paddy in the Brahmaputra Valley (Bezbaruah, 2006) and there are now clear indications of changes in the cropping pattern in the state. The present paper is induced by the need of a closer examination of the nature of cropping pattern and extent of crop diversification in the plains of Assam.

OBJECTIVES

The study is taken up with the following objectives.

(a) To examine the nature of cropping pattern in the plains of Assam.

(b) To investigate the extent of crop diversification across different districts in the plain areas of Assam.

(c) To identify the possible factors responsible for the observed variations in the crop diversification in the said areas.

Flood is a regular phenomenon in the state (sometimes twice or thrice in a year) causing a great havoc to the crop growing sector every year. An appropriate combination of crops may be used as a useful tool to minimize the possible devastations of flood[2]. Therefore, the central research question to be pursued in the present study is how and to what extent flood proneness influences the farmers' cropping pattern choice. That is, whether flood proneness induces or discourages diversification of crops.

The paper has been organized in five Sections. Section two pertains to a brief review of conceptualization of and the factors responsible for crop diversification whereas Section three deals with methodology and the model. While Section four covers the results and discussion Section five sums up conclusion and policy implications.

THEORETICAL BACKDROP

Conceptualization

Crop diversification can be broadly defined as a structural change in the crop choice in a particular geographical area. It refers to a change in the proportion of area (and other agricultural inputs) under different crops and a shift from one combination containing a few crops to another consisting of a larger number of crops.

Diversification of crops may take place either by a change in the proportion of area under each crop or a change in the value-wise contribution of each crop to total agricultural production. In the present study the first interpretation in terms of area has been used. However, a mere change in resources under different crops/crop groups does not necessarily mean crop diversification. For example, movements of resources from low value commodities mix to high value commodities mix may lead to specialization towards high value commodities rather than diversification of crops. Therefore, suitable diversification indices will be helpful to measure whether there is diversification or concentration of crops in a particular area.

There are several measures of crop diversification, which have been used extensively in the empirical studies (Pope and Prescott, 1980; Shiyani and Pandya, 1998; Gupta and Tewari, 1985) such as Herfindahl Index, Ogive Index, Entropy Index and Modified Entropy Index.

Herfindahl Index (H.I.):

$$H.I. = \Sigma P^2_i \qquad (i = 1,2, \ldots, N)$$

Where, N is the number of crops and P_i represents acreage proportion of the ith crop in total cropped area. With increase in diversification, the H.I. decreases.

Ogive Index (O.I):

$$O.I. = \Sigma \{P_i - (1/N)\}^2 \div (1/N). \qquad (i = 1, 2\ldots, N)$$

Like the H.I. the O.I. is a measure of concentration.

Entropy Index (E.I.):

$$E.I. = \Sigma P_i .\log(1/P_i). \qquad (i= 1, 2\ldots, N)$$

The E.I. increases with the increase in diversification and it approaches zero when there is perfect concentration. The upper bound of the index is log N.

Modified Entropy Index (M.E.I.):

The limitations of the Entropy Index have been sought to be overcome by Modified Entropy Index. In M.E.I. the base of the logarithm is taken as variable instead of a fixed base. It is computed as:

$$M.E.I. = \Sigma P_i .\log_N(1/Pi) \qquad (i= 1, 2\ldots, N)$$

Here the base of the logarithm is taken to be 'N' number of crops. This index has a lower limit equal to zero when there is complete concentration and it assumes upper limit of one when there complete diversification. It provides for uniformity and fixity to the scale used as norm to measure the degree of diversification unlike the E.I.

Each of these indices has its limitation. However, considering the relative superiority over the other measures the Modified Entropy Index has been used in the present study.

Factors Affecting Diversification of Crops

The available literature reflects upon various factors, which affect the nature and pace of crop diversification (Joshi et al, 2004; Singhal and Gauraha, 1997; Vyas, 1996; Gupta and Tewari, 1985; Anosike and Coughenour, 1990; Pope and Prescott, 1980).

Irrigation has a very significant impact upon diversification of crops. Singhal and Gauraha (1997), while studying crop diversification in Madhya Pradesh found that the factors responsible for crop diversification were mainly rainfall and irrigation. Likewise, Birthal et al (2006) found that irrigation mainly through tubewells or pumpsets has a positive and significant impact of the level of diversification. Similarly, in a study in the Brahmaputra Valley of Assam, Dutta et al (2005) found that the farmers have been able to diversify their cropping pattern after getting access to irrigation facilities, especially through privately owned shallow tube wells (STWs). On the contrary, Joshi et al (2004) found that diversification of crops in favour of horticultural commodities declines with increase in irrigated area.

Market plays a very crucial role in determining the pattern of crop diversification (Vyas, 1996). Among the market forces relative prices and profitability play a very significant role in this regard. Another important factor is market infrastructure in terms of better road connectivity and relative proximity to the market places. Urbanization, accompanied by changes in tastes and preferences, is also sometimes referred to as a factor influencing diversification of crops, particularly towards high value commodities (Rao *et al*, 2004).

The effect of size of the farm on diversification of crops has received due mention in several studies. Some of the studies (White and Irwin, 1972; Gupta and Tewari, 1985; Mishra *et al*, 2004) confirm a negative relationship between farm size and diversification. This may be due to the fact that one might expect the larger farms to be more specialized when there are scale economies in an enterprise. Another possible explanation is that since farm size and wealth tend to be positively correlated, one can deduce that wealthier farms are less risk averse and hence less diversified, other things being equal (Mishra *et al*, 2004). This is in sharp contrast with the findings of some other studies (Pope and Prescott, 1980; Anosike and Coughenour, 1990), which show a positive relationship between farm size and diversification. This may be due to absence of sufficient economies of scale in any commodity to warrant specialization for the large farms.

Attitude towards risk is also an important factor affecting the farmers' choice regarding crop combinations. Diversification has been widely held as an important risk mitigating strategy (Anosike and Coughenour, 1990; Barghouti et al, 2004). According to Shiyani and Pandya (1998) and Kumar et al (2002), crop diversification is suggested as an important strategy to avoid risk and uncertainty associated with agriculture due to climatic and biological vagaries.

METHODOLOGY AND THE MODEL

This paper is based on secondary data collected from different sources of Government of Assam such as Directorate of Agriculture, Planning and Development Department and Revenue Department. Data on area under different crops have been collected from the Directorate of Agriculture, Government of Assam. On the basis of this, proportion of area under ten major crops and crop groups have been calculated for the year 2003-04 across twenty one districts of Assam excluding the two hill districts of Karbi Anglong and North Cachar Hills because of the fact that the agricultural system in these two districts is different from that in the plain areas. Moreover, the research question of the study is not relevant in these two hill districts. The major crops and crop groups taken for the present study include the three types of paddy, namely winter paddy, summer paddy and autumn paddy[3], other cereals (maize, wheat and other cereals and small millets), total oilseeds, total fibre crops, total spices, total fruits, total vegetables and sugarcane. Winter paddy is the traditionally predominant paddy variety grown in Assam. However, due to availability of irrigation facilities in the recent years autumn paddy and summer paddy are also grown now a days. These two varieties have now assumed greater importance in providing stability of production of paddy in the state.

The proportions of area under these crops/crop groups broadly describe the cropping patterns across the districts. These proportions have further been summarized into a measure of crop diversification, viz., Modified Entropy Index (MEI).

To explain variations in the diversification of crops (measured by the M.E.I) across various districts a multiple regression model has been used. The available literature reflects on various factors, which affect the nature and pace of crop diversification as mentioned in section II.2. However, owing to non-availability and non-suitability of data on all these factors in the context of prevailing conditions in Assam only three important factors have been taken into account in the present study as possible influences responsible for the variations in the level of diversification across different districts. These are mentioned below.

Irrigation

Irrigation is often cited as a very important factor that affects the nature and speed of crop diversification in a particular area. However, the utilization rate of government irrigation schemes in Assam is very poor[4] (Dutta et al, 2005). The privately owned STWs are the major source of irrigation in the state, which is used more effectively than the government irrigation schemes. Therefore, irrigated area through STWs per hectare of net sown area (IRR) has been taken as an explanatory variable.

Agricultural Work Force

Diversification into a larger number of crops, particularly towards high value commodities requires more labour (Joshi *et al,* 2004). Thus, availability of agricultural workers may be an important determinant of diversification. Hence, agricultural work force (cultivators and agricultural labourers) per hectare of net sown area (AWF) has been taken as another explanatory variable.

Flood Proneness

In Assam one major source of risk and uncertainty that the farmers frequently[5] face is flood. Crop diversification is suggested as an important strategy to avoid risk and uncertainty associated with agriculture (Shiyani and Pandya, 1998; Kumar et al, 2002). Here data on cropped area affected by flood as proportion of net sown area would have been very useful to capture as to how and to what extent the uncertainties associated with flood influence the farmers' choice of cropping pattern and crop diversification. But due to non-availability of data for all the districts affected by flood the number of villages affected by flood as proportion to total number of villages (FP) has been taken as a proxy of flood proneness.

The degree of crop diversification has been calculated using MEI. To capture the effect of IRR, AWF and FP on diversification of crops across the plain districts of Assam the linear regression will not be suitable because of the value of diversification index being contained between 0 and 1. Therefore, a more appropriate functional form would be one in which the estimated value of the dependent variable will be contained within the range of 0 and 1. Hence, a logistic functional form would be suitable. The mathematical form of a logistic function is given by

$$Y = \frac{1}{1+e^{-Z}}$$

Where,

Y= Diversification index measured by MEI, (0d≤Yd≤1)

$z = \beta 0 + \beta_1 \text{ IRR} + \beta_2 \text{ AWF} + \beta_3 \text{FP}$

However, due to lack of the author's access to appropriate software for estimating the above non-linear function a transformation[6] of the function has been carried out as shown below.

$$Y = \frac{1}{1+e^{-Z}} = e^z / (1+e^z)$$

$$Y/(1\text{-}Y) = e^z$$

$$\text{Hence,} \quad \ln \frac{Y}{1-Y} = Z$$

$$\text{Let,} \quad \ln \frac{Y}{1-Y} = D$$

As $Y \to 0$, $D \to -\infty$
and $Y \to 1$, $D \to \infty$

Where D is a redefined measure of diversification but its value is not contained between 0 and 1. As the value of Y increases from 0 to 1, the value of D will range between $-\infty$ and ∞.

Now, adding a disturbance term the ultimate linear regression model formulated for the present study is

$$D = \beta_0 + \beta_1\, IRR + \beta_2\, AWF + \beta_3 FP + u$$

Where, u is the random disturbance term that is assumed to fulfill the classical standard least square assumptions and β_0, β_1, β_2 and β_3 are the parameters to be estimated.

RESULTS AND DISCUSSION

The cropping pattern and the extent of crop diversification across the different plain districts of Assam can be seen from Table 1. The districts are classified into three broad categories, viz., highly diversified (M.E.I.e ≥ 0.7), moderately diversified ($0.7 >$M.E.I.e ≥ 0.5) and less diversified (M.E.I < 0.5). If we look at the extent of crop diversification (measured by M.E.I.) as portrayed in Table 1, we see that the area comprising of Dhubri, Darrang, Barpeta, Goalpara, Nagaon, Bangaigaon, Kamrup, Marigaon, Kokrajhar is highly diversified; the area comprising of Sonitpur, Nalbari, Lakhimpur, Dhemaji, Tinsukia, Golaghat and Cachar is moderately diversified whereas the area consisting of Hailakandi, Jorhat, Dibrugarh, Karimganj and Sibsagar is less diversified. It is interesting to note that all the highly diversified districts follow a particular geographical pattern and lie in the Lower and Central Brahmaputra Valley.

The results of the regression analysis of crop diversification are presented in Table 2. The model gives a reasonably good fit in terms of R^2 value of 0.69. The F statistic is significant at

Table 1
Proportion of Area under Different Crops and Modified Entropy Index of Diversification of Crops across Districts of Assam for the Year 2003-04

District	*Autumn Rice*	*Winter Rice*	*Summer Rice*	*Other Cereals*	*Total Oilseeds*	*Total Fibre*	*Total Spices*	*Total Fruits*	*Total Vegetables*	*Sugarcane*	*M.E.I*
Highly Diversified											
Dhubri	0.13	0.27	0.25	0.08	0.08	0.08	0.04	0.01	0.06	0.00	0.83
Darrang	0.25	0.36	0.06	0.03	0.13	0.05	0.03	0.02	0.08	0.00	0.78
Barpeta	0.22	0.38	0.07	0.04	0.13	0.03	0.03	0.02	0.08	0.00	0.77
Goalpara	0.17	0.48	0.14	0.00	0.05	0.03	0.03	0.04	0.06	0.00	0.76
Nagaon	0.12	0.44	0.17	0.02	0.09	0.03	0.02	0.03	0.06	0.02	0.76
Bongaigaon	0.24	0.41	0.07	0.05	0.10	0.03	0.02	0.03	0.06	0.00	0.75
Kamrup	0.13	0.44	0.18	0.03	0.09	0.02	0.03	0.04	0.05	0.00	0.74
Morigaon	0.10	0.34	0.34	0.04	0.06	0.03	0.03	0.02	0.02	0.01	0.73
Kokrajhar	0.20	0.44	0.06	0.03	0.15	0.03	0.02	0.02	0.05	0.00	0.71
Moderately Diversified											
Sonitpur	0.15	0.54	0.06	0.02	0.10	0.01	0.02	0.03	0.06	0.01	0.67
Nalbari	0.11	0.57	0.09	0.02	0.08	0.01	0.02	0.02	0.08	0.00	0.64
Lakhimpur	0.11	0.55	0.12	0.01	0.12	0.00	0.02	0.03	0.05	0.00	0.64
Dhemaji	0.13	0.55	0.08	0.03	0.11	0.00	0.02	0.03	0.04	0.00	0.64
Tinsukia	0.08	0.65	0.00	0.01	0.10	0.00	0.02	0.05	0.08	0.00	0.53
Golaghat	0.04	0.69	0.03	0.02	0.03	0.01	0.02	0.04	0.10	0.03	0.53
Cachar	0.07	0.68	0.07	0.00	0.02	0.00	0.01	0.07	0.08	0.00	0.51
Low Diversified											
Hailakandi	0.11	0.70	0.04	0.00	0.02	0.00	0.02	0.04	0.07	0.00	0.49
Jorhat	0.03	0.73	0.01	0.01	0.07	0.00	0.02	0.04	0.08	0.00	0.46
Dibrugarh	0.05	0.72	0.00	0.01	0.11	0.00	0.02	0.03	0.06	0.00	0.45
Karimganj	0.05	0.77	0.05	0.00	0.00	0.00	0.02	0.04	0.06	0.01	0.41
Sibsagar	0.01	0.88	0.00	0.00	0.01	0.00	0.02	0.03	0.04	0.00	0.25

0.01 level which shows the overall significance of the estimated relation. The coefficient of IRR is positive and significant at 0.01 level which implies that irrigation through STWs has a positive and very significant effect on crop diversification. Similarly, the coefficient of AWF is positive and significant at 0.05 level. It implies that greater availability of AWF promotes diversification of crops. On the other hand, the coefficient of FP is negative and significant at 0.10 level which implies that flood proneness reduces scope for diversification.

Table 2
Results of Regression Analysis of Crop Diversification Across Districts of Assam

Variables/Particulars	*Estimated Coefficients/Values*
Irrigated area through STWs per hectare of Net Sown Area(IRR)	6.61***
	(1.08)
Agricultural Work Force per hectare of Net Sown Area(AWF)	0.49**
	(0.18)
Proportion of villages affected by flood (FP)	-1.20*
	(0.60)
Constant	-.97
	(0.35)
R^2	0.69
F (degree of freedom =3, 17)	12.90***

Notes: Figures in parentheses represent standard errors of the respective coefficients.
*, ** and *** indicate significant at 0.10, 0.05 and 0.01 level.

CONCLUSION AND POLICY IMPLICATIONS

The study reveals that irrigation through STWs and availability of agricultural work force have positive and significant role in diversification of crops in the plains of Assam. On the other hand, flood proneness restricts crop diversification. This may be due to the possible reason that the cropping season in the flood prone areas is limited by flood. In the remaining part of the year the farmers may be forced to confine to staple crops and hence, the scope for diversification of crops to high value commodities gets restricted. However, this aspect requires further investigation at the micro level.

The policy implication which arises from the study is that adequate steps should be taken to develop the irrigation infrastructure in the state to facilitate diversification of crops. In view of the greater effectiveness of STWs based privately owned small scale irrigation system in the state private investment should be encouraged in such system (Dutta et al, 2005). At the same time, considering a very low rate of utilization of the existing potential under government irrigation schemes[7], appropriate steps should be taken to make such potential more effective.

NOTES

1. About half the farmers in the state cultivate and harvest their field only once a year (Government of Assam, 2003).
2. Purkayastha (2005) has shown how the farmers of a village in the Nagaon district of Assam are trying to cope up with the recurrent and prolonged floods and experiment with different crop combinations, often successfully in low lying fields or heavily silted land
3. Winter paddy (Sali), autumn paddy (Ahu) and summer paddy (Boro) are so named on the basis of their harvesting periods. However, the harvesting periods of these varieties have now been pre-poned to a great extent following the introduction of short duration HYV seeds. Autumn paddy is now harvested before the onset of monsoon, summer paddy in spring and winter paddy in the end of autumn.
4. The utilization rate of the government irrigation schemes in the state in 2002-03 was only 15.26 per cent of the total created potential (Government of Assam, 2004)
5. sometimes twice or thrice a year
6. This is sometimes referred to as logit transformation. A similar transformation was used by Mishra *et al* (2004).
7. The utilization rate of the government irrigation schemes in the state in 2002-03 was only 15.26 per cent of the total created potential (Government of Assam, 2004).

REFERENCES

Anosike, Nnamdi and C. Milton Coughenour, (1990), "The Socioeconomic Basis of Farm Enterprise Diversification Decisions", *Rural Sociology*, Vol. 55, No. 1.

Barghouti, Shawki, Samuel Kane, Kristina Sorby and Mubarik Ali (2004), *Agricultural Diversification for the Poor: Guidelines for Practitioners*, Agriculture and Rural Development Discussion Paper 1, The World Bank, Washington, D.C., U.S.A.

Bezbaruah, M.P. (2006), "Rapporteurs' Report on 'Agricultural Development in the North-East: Status, Assessment and Prospects", *Indian Journal of Agricultural Economics*, Vol. 61, No. 3.

Birthal, P. S., A. K. Jha, P. K. Joshi and D. K. Singh (2006), "Agricultural Diversification in North Eastern Region of India: Implications for Growth and Equity", *Indian Journal of Agricultural Economics*, Vol. 61, No. 3.

Datta, Samar K. and R.L. Shiyani (2005), "Agriculture-Changing Paradigms", *Yojana*, Ministry of Information and Broadcasting, New Delhi, Vol. 49, January.

Dutta, M. K., Raju Mandal and M. P. Bezbaruah (2005), "Crop Diversification under Government Supplied and Privately Owned Irrigation: A Study in the Brahmaputra Valley in Eastern India", *Assam Economic Journal*, Vol. XVIII.

Government of Assam (2003), *Assam Human Development Report 2003*, Planning and Development Department, Dispur, Guwahati.

————(2004), *Economic Survey, 2003-2004*, Directorate of Economics and Statistics, Dispur, Guwahati

————(2005), *Handbook of Agricultural Statistics, 2004-05*, Directorate of Agriculture, Khanapara, Guwahati.

Government of India (2006), *Towards Faster and More Inclusive Growth: An Approach to the 11th Five Year Plan*, Planning Commission, New Delhi.

Gupta R. P. and S. K. Tewari (1985), "Factors Affecting Crop Diversification: An Empirical Analysis", *Indian Journal of Agricultural Economics*, Vol. 50, No. 3.

Joshi, P. K., A. Gulati, P. S. Birthal and L. Tewari (2004), "Agricultural Diversification in South Asia: Patterns, Determinants and Policy Implications", *Economic and Political Weekly*, Vol. 39, No. 24.

Kumar, Ashok, S. K. Sharma and G. D. Vashist (2002), "Profitability, Risk and Diversification in Mountain Agriculture: Some Policy Issues for Slow Growth Crops", *Indian Journal of Agricultural Economics*, Vol. 57, No. 3.

Mishra, Ashok K., Hisham S. El-Osta and Carmen L. Sandretto (2004), "Factors Affecting Farm Enterprise Diversification", *Agricultural Finance Review*, Fall.

Pope, Rulon D. and Richard Prescott (1980), "Diversification in Relation to Farm Size and other Socioeconomic Characteristics", *American Journal of Agricultural Economics*, Vol. 62, No. 3.

Purkayastha, Gautam (2005), "Benegati—A Model Agricultural Village", *Kurukshetra*, Vol. 53, No. 4.

Rao, Parthasarathy, P. S. Birthal, P.K. Joshi and D. Kar (2004), *Agricultural Diversification in India and Role of Urbanization*, MTID Discussion Paper No. 77, Market, Trade and Institutions Division, International Food Policy Research Institute, Washington, D.C., U.S.A.

Shiyani, R. L. and H. R. Pandya (1998), "Diversification of Agriculture in Gujarat: A Spatio-Temporal Analysis", *Indian Journal of Agricultural Economics*, Vol. 53, No. 4.

Sidhu, R. S. and M.S. Dhillon (1997), "Land and Water Resources in Punjab: Their Degradation and Technologies for Sustainable Use", *Indian Journal of Agricultural Economics*, Vol. 52, No. 3.

Sidhu, J.S. and R.S. Sidhu (1988), "Growth and Area Response of Commercial Crops in Punjab: Some Policy Issues", *Indian Journal of Agricultural Economics*, Vol. 43, No. 3.

Singhal, M. and A. K. Gauraha (1997), "Crop Diversification in Madhya Pradesh: An Economic Analysis", Paper Presented at the Annual Conference of Indian Society of Agricultural Economics.

Vyas, (1996), "Diversification in Agriculture: Concept, Rationale and Approaches", *Indian Journal of Agricultural Economics*, Vol. 51, No. 4.

8

DYNAMICS OF CROPPING PATTERN IN KERALA–SPATIAL INTERTEMPORAL ANALYSIS

K. P. Mani

INTRODUCTION

Current free trade strategy in India has necessitated significant changes in the cropping pattern away from coarse cereals, pulses, oil seeds and sugarcane in favour of superior cereal, horticultural crops, vegetables and livestock products. These changes are bound to have important effects on inter regional, regional and inter clan equity among country's farmers. Major among them is the threat against the food security of the country. "The fears about food security are not without a firm basis but is amply borne out by the experience of several African and Latin American countries which have adopted the policy of relying on export of fruits, flowers and vegetables to the neglect of domestic food grain economy, an experience which has been well documented"(Ajithkumar Singh 1996).

In India, just after green revolution gross area under cultivation remained at 1578 lakh hectares, of which the contributions of food crops and non food crops respectively being 78.78 per cent and 21.22 per cent. All the states except Kerala and Gujarat had reasonable area under food crops. But compared to these trends, the area under food crops substantially declined since 1985. In 1992-93, gross cropped area remained at 1854 lakh hectares, the shares of food crops and non food crops respectively being 67.24 per cent and 25.27 per cent. Though, the fall in the share of area under food crops is marginal compared to the hares in 1970-71, many states recorded significant reduction in area under food crops. The states among this category are, Andhra Pradesh, Gujarath, Haryana, Karnataka, Kerala, Orissa, Rajasthan and Tamil Nadu.

SIGNIFICANCE OF THE STUDY

The reduction in area under food crops in Kerala from 40.43 per cent in 1970-71 to 18.74 per cent in 1992-93 and 16.52 in 2002-03, is a phenomenon happened very rarely in any country. The gravity of the problem is more severe when the state is 50 per cent short of food grains. The area index of food grains has lost considerably and continuously from 102.55 in 1976-79 to 58.38 in 1999-02, whereas the gain in the productivity was not adequate enough to make good this short fall. On the other hand, the index of area under non food crops increased from 100.52 in 1976-79 to 159 in 1999-02. It is also worthwhile noting that among the non food categories, the growth of area expansion index is tremendous in the case of coconut and rubber. Hence in this paper an attempt is made to examine the spatial and inter temporal shifts in cropping pattern in Kerala and the determinants causing this shift.

METHODOLOGY

The study is based on secondary data collected for the period 1975-76 to 2002-03. To reveal inter temporal shifts, the entire period was disaggregated to decennial data and compound growth rates were calculated with the support of exponential functions and kinked exponentials.

$Y = \infty^{\rho}$, $\text{Log } Y = \log \infty + t \log^{\rho}$, CGR = anti log $(1-\rho) \times 100$, where Y = relevant variable.

The motive factors were analysed theoretically using supply response models, because in agriculture, crop shift decisions and resultantly supply decisions are taken months before the actual products reach the market and prices are realised. The production decisions are therefore taken on the basis of expected prices. Accordingly,

Different models have been developed, the major being,

Cobweb Model

The commonly used proxy for expected price is the previous year's actual price,

$$P_t = P_{t-1}$$

Adaptive Expectations

This formulation too supposes that price expectation is based on past price or prices, but the agents are allowed to learn from mistakes so that expected price is given by,

$$P_t = P_{t-1} + (P_t - P_{t-1)}$$

Nerlovian Adjustment Log Model

Model I

$$A_t = b_0 + b_1 RP_{t-1} + b_2 R\,Y_{t-1} + b_3\, CV_{RP} + b_4\, CV_{RY} + b_5 A_{t-1} + V_t$$

$$A_t = b_0 + b_1 RPR_{t-1} + b_2\, CV_{RPR} + b_3 I_t + b_4\, SSR_t + b_5\, A_{t-1} + v_t$$

Where,

At = the actual area planted and hectares under the crop concerned

RP_{t-1} = relative price in terms of competing crop

RY_{t-1} = relative yield in terms of competing crop

CV_{RP} CV_{RY} coefficient of variation of the preceding three years relative price And yield

RPR_{t-1} = relative gross profitability in terms of competing crop

CV_{RPR} = a measure of relative profitability risk

I_t = gross irrigated area under all crops

SSR_t = rainfall of sowing season for the concerned crop

Arima Process

The ARIMA process expectation also assumes that past price determines price expectations but past disturbances are taken into account as well. Here,

$P_t = P_t - e_t = \prod P_{t-1} + \prod P_{t-2} + \prod P_{t-3} + \ldots\ldots\ldots\ldots\ldots\ldots + \prod P_{t-q} + \partial e_{t-1} + \partial e_{t-2} + \ldots\ldots\ldots\ldots\ldots\ldots\ldots\ldots\ldots\ldots\ldots$

THE MODEL AND ESTIMATION

The above mentioned models cannot be as such applied for the Kerala agriculture because of the weak data base. Hence, the following models were developed and estimated considering problems of data deficiency and quality

Area response model,

$$A_t = A_0 + a_1 P_{t-1} + a_2 R_t + a_3 W_t + V_t \quad (1)$$

$$(A_t - A_{t-1}) = \partial\,(A_t - A_{t-1}) \quad (2)$$

Yield response model

$$Y_t = a_0 + a_1\, P_{t-1} + a_2 R_t + a_3\, W_t + V_t \quad (3)$$

$$(Y_t - Y_{t-1}) = \partial\,(Y_t - Y_{t-1}) \quad (4)$$

Where,

A_t = desired area in hectares under the crop in the year t

A_{t-1} = actual area in hectares under the crop in the year t

P_t = farm harvested price in the period t

R_t = price risk as measured by the standard deviation of price in the past 3 Years from the period t

W_t = pre sowing months average rainfall in the period t

W = average rainfall in the year t

Y_t = actual yield of crop in kgs in the year t

∂ = coefficient of adjustment

DISCUSSION

Both geographically and climatically Kerala forms a distinct region, whose topography is dominated by the Western Ghats. Kerala is one of the states in India, where its land resources are put to more intensive use than any where else mainly because of the low per capita availability of land in the state.

According to the current statistics (2005) out of the total geographical area of 3885 thousand hectares, one fourth was under forests and about 1.26 per cent was barren and uncultivable land. Another 8.28 per cent was put to non agricultural use. Kerala had a net sown area of 2239 thousand hectares in 2001-02 forming 57.63 per cent of the total area. The area sown more than once was just 808 thousand hectares compared to 3123 hectares in 1952-53, the intensity of cropping being 1.36.

Out of the gross cropped area of 2349 thousand hectares in 1960-61, rice occupied 778.91 hectare, forming 33.15 per cent of the gross cropped area. But the share of gross cropped area under rice consistently came down to 28 per cent in 1980-81 and 15.45 per cent in 2002-03. In absolute terms, the area under rice declined from 778.91 thousand hectares in 1960-61 to 678.28 thousand hectares in 1980-81, the average annual fall being 0.68 per cent. Further between 1980-81 and 1999-00, the area under rice declined by 30.53 per cent, absolute fall during the 15 years period being 207.13 thousand hectares. The fall in cumulative area in 1995-96, compared to 1960-61 being 307.76 thousand hectors. The area divided for all food crops together was about 46 per cent in 1960-61 but it came down to just 16.52 per cent in 2002-03. The major reason is the domination of cash crops like coconut and rubber. In the year 1960-61, coconut constituted 21.31 per cent and rubber constituted 5.23 per cent of the gross cropped area. These shares marginally improved to 22.77 per cent and 8.24 per cent respectively by 1980-81. But since 1980-81, there is a tremendous leap. The area under coconut reached 982 .10 in 1995-96, almost double the area in 1960-91. Similarly, area under rubber reached 449 thousand hectares in 1995-96, an increase of 3.68 times compared to 1960-61. While the area under rice declined by 336.55 thousand hectares between 1980-81 and 1999-00, area under coconut and rubber increased by 330.73 thousand and 211.23 thousand during the same period.These three crops constituted 60 per cent of the gross cropped area in 1960-61 and 62 per cent in 1999-00 and the trends clearly establish the claim that Kerala farmers are shifting the area under rice to coconut and rubber while the state is 50 per cent short of rice production compared to consumption requirements and when the state lacks linkages to absorb coconut and rubber.

In the succeeding section an attempt is made to examine the inter district, intra district and inter temporal shifts in cropping pattern with reference to rice, coconut and rubber on production and yield.

CROPPING PATTERN: INTER DISTRICT ANALYSIS

Inter district analysis presented in Tables 1 and 4 clearly show that area under rice declined in all the districts between 1975-76 and 2002-03. Maximum fall occurred in Trichur district (64.28 thousand hectares), followed by Kozhikode, Malappuram and Palakkad districts. There is fall in Kannur and Alappuzha districts also, but it is due to the division of these districts respectively to Kasaragod and Pathanamthitta. All the districts together made a decline of 382.27 hectares during the time span. Decade wise, the fall is 242.97 thousand hectares between 1975-76-1985-86 and 139.30 between 1985-86 -1999-00. Hence, we may infer that the relative contribution of rice to the total cropped area consistently declined in all the districts in the state, the steep fall being in northern parts.

Naturally, this fall should be reflected in an increase of the area of other crops particularly coconut and rubber felt "Apriori". In the case of area under coconut, maximum increase is recorded in northern districts, particularly Malappuram (32.78 thousand hect) and Kozhikode (22.16 thousand hect) between 1975-76 and 2002-03. During the same period Trichur district and Palakkad district respectively recorded an increase of 23.89 thousand hectares and 26.70 thousand hectares. It is important to note that Kollam and Kottayam districts record a fall in the area under coconut. During the period 1975-76 to199-00 the net increase in the area under coconut in the state is 102.06 thousand hectares, decade wise split being -50.37 thousand hectares and 152.43 thousand hectares. Compared to the fall in the area under rice between 1985-86 and 1999-00 (139.30thousand hect) it is seen that the spread in area under coconut is higher than the spread in the case of staple food rice. Hence we may infer that, in the recent

Table 1

Quinquennial and Decennial Changes in the Area, Production and Yield of Rice in Kerala (1975-76 to 2002-03)

Districts	(A) 75/76 85/86	(A) 85/86 90/91	(A) 90/91 95/96-	(A) 95/96 02/03	(P) 75/76 85/86	(P) 85/86 9091	(P) 90/91 95/96	(P) 95/96 02/03	(Y) 75/76 85/86	(Y) 85/86 90/91	(Y) 90/91 -95/96	(Y) 95/96 02/03
TVM	-11.1	-4.67	-4.33	-4.22	-11.9	-8.7	-8.1	-7.2	244	-18	-27	-23
KLM	-18.2	-4.28	-5.23	-4.65	-20.8	-2.45	-11.5	-10.4	208	165	-61	-48
ALP	-40.2	-4.63	-5.81	-4.87	-28.9	-19.6	-32.4	-31.5	536	172	361	369
KOT	-12.2	-5.59	-1.25	-0.98	-17.9	-4.61	-6.61	-2.36	100	-567	-165	-89
IDU	-7.6	3.17	-0.38	-0.12	-9.3	-5.8	-0.85	-0.41	395	118	-10	-05
EKM	-14.2	-21.7	-3.06	-4.01	10.22	-40.1	-1.64	-0.24	345	-55	56	28
THR	-31.2	-21.1	-11.8	-10.5	-10.2	-22.6	-15.2	-11.4	313	150	89	74
PGT	-13.4	-15.1	-5.6	-2.8	-42.6	17.9	-11.1	-8.65	-98	322	10	8
MAP	-23.4	-13.5	-9.8	-4.5	-32.1	-12.2	-10	-8	15	134	128	121
KOZ	-37.3	-6.6	-8.6	-5.3	-43.2	8.1	-4.2	-2.1	14	36	17	21
KNR	-33.9	-31.0	-22.8	-20.1	33.5	-46.2	-6.5	-5.4	231	98	-143	-141

Area in 000 hectares production in 000 metric tonnes
TVM Thiruvanathapuram KLM Kollam ALP Alapuzha KOT Kottayam IDU Idukki EKM Ernakulam
THR Thrissur PGT Palakkad MAP Malappuram KOZ Kozhikode KNR Kannur 3 districts are not covered in this study

Source: (1) Government of Kerala (1975) Agricultural Statistics in Kerala, The Bureau of Economics and Statistics, Thiruvanathapuram
(2) Government of Kerala, Economic Review (various years), State Planning Board, Thiruvanathapuram

Table 2

Quinquennial and Decennial Changes in the Area, Production and Yield of Coconut in Kerala 1975-76 to 2002-03

Districts	*(A) 75/76 85/86*	*(A) 85/86 90/91*	*(A) 90/91 95/96-*	*(A) 95/96 02/03*	*(P) 75/76 85/86*	*(P) 85/86 9091*	*(P) 90/91 95/96*	*(P) 95/96 02/03*	*(Y) 75/76 85/86*	*(Y) 85/86 90/91*	*(Y) 90/91- 95/96*	*(Y) 95/96 02/03*
TVM	0.98	12.5	2.39	2.47	-.074	.099	.134	.135	-1.56	1.80	.65	.64
KLM	-29.3	9.10	1.21	1.34	-.141	.020	.033	0.187	1.32	.890	.674	.587
ALP	-24.1	17.9	-2.1	-1.8	.110	–	–	–	1.13	.272	-.077	-0.74
KOT	-0.91	-1.81	-5.68	-3.21	.100	.007	–	–	-1.72	1.24	.43	.41
IDU	6.11	2.72	1.51	1.24	-0.01	0.02	0.015	0.014	-2.85	1.56	0.798	0.478
EKM	8.90	6.63	-2.36	-1.54	.058	.019	.067	.047	-.515	1.26	.400	.487
THR	9.66	20.4	6.26	5.47	.048	.018	.122	.154	.173	.589	.087	.047
PGT	9.35	11.80	5.55	5.21	.024	.050	.051	.024	-.701	.600	.966	.687
MAP	-5.03	37.1	.812	.784	.047	.192	.147	-.147	-.213	.732	.888	.784
KOZ	14.0	10.5	-2.4	-2.57	-.067	.188	.170	.174	-.812	1.07	1.17	1.54
KNR	31.8	31.5	6.85	5.68	-.001	.040	.034	.035	-0.03	.044	.024	.014

Area in 000 hectares Production in 000 metric tonnes

Source: (1) Government of Kerala (1975) Agricultural Statistics in Kerala, The Bureau of Economics and Statistics, Thiruvanathapuram

(2) Government of Kerala, Economic Review (various years), State Planning Board, Thiruvanathapuram

years, there is definitely a trend in favour of coconut plantations in Northern Kerala.

Kerala state contributed nearly 90 per cent of rubber production in the country. Inter district data reveal that since 1975-76, rubber plantations were extended to all the districts. It is also seen that over the last 25 years, area under rubber cumulatively increased by 169.13 thousand hectares.

In the succeeding paragraphs an attempt is made to examine the variations in production and yield with respect to rice, coconut and at the inter district level during the period 1975-76 to 2002-03. Rice production improved from 1067.53 thousand tonnes in 1960-61 to 1364.87 thousand tonnes in 1975-76. Since then it started to decline and the current production (2003) is 1187.2 thousand tonnes. The productivity being 1558kg/hect in 1975-76 and 2012kg/hect in 2002-03. The districts which recorded maximum in production are, Palakkad and Thrissur districts of which Palakkad district served as the "rice bowl" of Kerala immediately after the formation of the state in 1956.

The production of coconut increased from 3439 million nuts in 1975-76 to 5303 million nuts in 1999-00, respective yield being 4962 and 5890. Among the districts, net additions are the maximum in Kozhikode, Malappuram and Kannur districts followed by Thrissur, Palakkad, Ernakulam and Thiruvanathapuram districts. It is striking to observe that in Kollam district production came down.

Kerala has the virtual monopoly over production of rubber in the country. The state's rubber production increased by 3.14 times and yield 1.63 times over the last 20 years. In absolute terms, yield improved from 611kg/hect in 1975-76 to 1000 in 1999-00. Leaving home district Kottayam, the other districts made commendable performance of production of rubber are Ernakulam, Thrissur and Thiruvanathapuram. The good performance of production and yield is definitely due to the efforts taken by the Rubber Board. But the crucial question as raised earlier is whether the excess supplies of rubber in relation to consumption "within state" will able to find attract sufficient margins permanently. It is true that still there is wide

Table 3

Quinquennial and Decennial Changes in the Area, Production and Yield of Rubber in Kerala 1975-76 to 2002-03

Districts	*(A) 75/76 85/86*	*(A) 85/86 90/91*	*(A) 90/91 95/96-*	*(A) 95/96 02/03*	*(P) 75/76 85/86*	*(P) 85/86 9091*	*(P) 90/91 95/96*	*(P) 95/96 02/03*	*(Y) 75/76 85/86*	*(Y) 85/86 90/91*	*(Y) 90/91- 95/96*	*(Y) 95/96 02/03*
TVM	100	4.01	3.51	3.57	3.53	7.03	9.92	8.54	-.180	.231	.290	.357
KLM	-6.4	4.54	2.33	2.98	-4.37	7.14	9.82	8.35	-.	.017	-.027	-.214
ALP	-.678	.114	.193	.198	.313	-.582	.833	.874	.226	-.202	.200	.201
KOT	28.37	79.37	-55.45	-10.2	16.11	32.71	30.37	28.65	–	.205	.243	.547
IDU	15.19	4.46	1.49	1.24	.054	11.02	9.90	8.97	-.139	.292	-.196	.214
EKM	17.71	12.38	3.64	3.57	7.12	18.16	20.62	18.74	-.082	.234	.327	.354
THR	1.27	.877	1.22	1.34	-11.1	3.02	4.68	2.47	-.203	.223	.293	2.96
PGT	7.73	4.38	4.62	5.47	3.33	5.38	7.88	6.54	–	.167	.201	.187
MAP	4.03	1.21	3.42	2.41	2.28	3.43	7.47	6.54	–	.123	.200	2.01
KOZ	-2.19	1.68	.844	.748	-1.47	4.16	5.35	4.57	–	.192	.263	265
KNR	-1.43	4.01	4.29	4.17	-.697	7.26	7.69	7.41	–	.208	.167	.147

A area in 000 hectares Production in 000 metric tonnes

Source: (1) Government of Kerala (1975) Agricultural Statistics in Kerala, The Bureau of Economics and Statistics, Thiruvanathapuram

(2) Government of Kerala, Economic Review (various years), State Planning Board, Thiruvanathapuram

gap between present production and consumption. But the relevant question is whether the gap will be filled by domestic rubber or imported rubber.

The above discussions validate our hypothesis that the cropping pattern of Kerala state has turned in favour of cash crops particularly coconut and rubber at the cost of area under rice, the staple food of the state.

Table 4
Gross Change in Area between 1975-2003 (000 hect)

Districts	*Rice*	*Coconut*	*Rubber*	*Gap*	*Gap after 2% for other crops*
Thiruvanathapuram	-20.11	15.86	17.62	–	–
Kollam	-27.77	-18.98	0.45	46.30	37.04
Alapuzha	-41.45	-8.35	-0.371	50.17	40.14
Kottayam	-19.12	-16.63	52.30	–	–
Idukki	-11.18	4.90	21.15	–	–
Ernakulam	38.99	13.17	33.73	–	–
Thrissur	64.28	23.89	11.26	12.63	10.10
Palakkad	34.21	26.70	16.74	–	–
Malappuram	46.80	32.78	8.67	5.35	4.28
Kozhikode	52.65	22.16	0.34	30.15	24.12
Kannur	67.16	6.38	6.87	53.91	43.12
Total	382.27	102.06	169.13		158.80

Computed from Tables 1, 2 and 3.

Determinants From Table 5, it is seen that the coefficient of determination ranged from 0.2950 and 0.9861 in the districts under study. The significance of R square values were indicative of the fact that the variables included in the area responses models were capable of explaining sizable proportion of the variation in the area. The coefficient of farm price was positive and significant in six districts, namely Thiuvanathapuram, Alappuzha, Ernakualam, Trichur, kozhikode and Kannur. On the other hand the coefficient remained to be negative and insignificant in three districts. Also it is surprising to note that in Palakkad district, one of the major rice producing centres of the state, the coefficient seems to be negative and insignificant. Hence, we may infer that generally area under rice in Kerala is price responsive.

The average rainfall was used as a proxy for computing the influence of weather on area allocation decisions. Its effect was found to be positive in seven districts out of ten, but the

Table 5
District wise estimated Regression Coefficients of the Price Elasticities of Area of Rice

Dist	*R Square*	*∞*	*Coef P_{t-1}*	*Coef A_{t-1}*	*Coef R_t*	*Coef W_t*	*Elsas. SRE*	*Elsas. LRE*
Thiruva Nathapur	0.513**	0.823	0.031**	0.13	-0.04*	0.0007	0.516	0.593
Alapuzha	0.45	1.79	0.009***	0.249	-0.034**	0.0009*	0.33	0.44
Kollam	0.55	12.89	-0.007	0.67**	0.11*	0.006	-0.01	-0.04
Kottayam	0.61*	42.56	0.122	0.507**	-0.02	-0.009	0.011	0.02
Ernakulam	0.29	1.28	0.003*	0.038	-0.005	0.0001	0.17	0.17
Thrissur	0.74	0.68	0.003**	0.47	-0.005	-0.0001	0.16	0.31
Palakkad	0.98***	116.17	-0.07	0.85**	-0.02	-0.003	-0.03	-0.25
Malappuram	0.40	36.81	-0.11	0.51***	0.32	0.03	-0.16	0.33
Kozhikode	0.60	0.72	0.008***	0.27	0.03	0.0001	0.33	0.46
Kannur	0.52	0.33	0.12*	0.66**	0.23	0.04	0.28	0.84
Kerala	0.24	914.62	0.12	0.28	0.57	0.04	0.01	0.014

*significant at probability level 0.1
**significant at probability level 0.05
***significant at probability level 0.01

coefficient was found to be significant only in Alappuzha. However, this inference supports our finding that the rice cultivation in Kerala is rainfed. In other words, we may conclude that the rainfall favourably influenced the area allocation decisions of the farmers. The risk factor which is represented by the standard deviation of price of immediately preceding three years was found to be negative in seven districts indicating farmers risk aversion behaviour. This may be one of the reasons for the claim by the farmers that rice cultivation is uneconomic in Kerala and hence they switch over to the cash crops.

From Table 6, it is seen that coefficient of determination was significant in Thiruvanathapuram, Alappuzha, Kollam, Trichur, Palakkad and Kozhikode. The farm harvest price was found to be significant only in two districts namely, Trichur and Kozhikode. Hence, we may infer that the price factor is not playing a significant role in deciding the yield. The

Table 6
District wise Estimated Regression Coefficients of the Determinants and Price Elasticities of Yield of Rice

Dist	*R Square*	*∞*	*Coef P_{t-1}*	*Coef A_{t-1}*	*Coef R_t*	*Coef W_t*	*Elsas. SRE*	*Elsas. LRE*
Thiruva Nathapur	0.79**	0.38	0.03	0.62***	-0.04	0.11***	0.04	–
Alapuzha	0.36**	0.38	0.03	0.02	0.07	0.14***	0.02	0.11
Kollam	0.78***	0.81	0.18	0.62***	-0.43	0.03	0.08	0.03
Kottayam	0.86***	0.80	0.01	0.86	0.01	0.01	-0.02	0.23
Ernakulam	0.29	0.53	0.04	0.21	0.02	0.01	0.05	0.06
Thrissur	0.75***	0.17	0.01	0.27	-0.02	-0.001*	0.35	0.49
Palakkad	0.67***	0.65	0.01	0.21	0.01	0.01	0.28	0.19
Malappuram	0.29	0.80	0.88	0.34*	0.04	0.001	0.27	0.42
Kozhikode	0.57**	0.61	0.01***	0.20	0.03	0.001	0.51	0.59
Kannur	0.14	0.79	0.003	0.21	0.014	0.0007	0.06	0.06
Kerala	0.56	0.64	0.006**	0.05	-0.02	0.004	0.36	0.38

*significant at probability level 0.1
**significant at probability level 0.05
***significant at probability level 0.01

inference is very relevant for Kerala because, the farmers are reluctant to improve yield by applying more inputs primarily due to the disproportionate growth of input prices compared to stagnant output price. The price risk factor was negative in many districts again reflecting the farmers risk aversion behaviour as revealed in the case of area response. Hence, we may infer that the situation of rice production in the state can be augmented only if some policy prescriptions are launched by the government to make the farmers risk bearers. The rainfall variable was found to have a positive impact on productivity levels in all districts except Ernakulam. Thus we may also suggest that the yield can be improved by adopting better technology involving adequate use of inputs such as fertilizers, HYV seeds.

CONCLUSION

Shifts in cropping pattern in favour of cash crops in majority of economies in recent years and the consequent reduction in area under food crops weakens the sustainability of food availability. The Indian economy characterised by low per capita income and large population would bring a rapid growth in the direct and derived demand for food grains. The reduction in area under food crops in Kerala from 40.43 per cent in 1970-71 to 18.74 per cent in 1992-93 and 16.52 in 2002-03, is a phenomenon occurred very rarely in any country. The study clearly established the claim that Kerala farmers are shifting the area under rice to coconut and rubber while the state is 50 per cent short of rice production compared to requirement. The study also revealed that, district wise, the crop shifts are more in Northern Kerala and the two crops towards which there is a significant shift are rubber and coconut displacing rice, in conformity with the state level trend. The farmers supply responses were also examined with the help of supply response models. Supply response estimates helped to identify the major factors deciding farmer responses, the major being, farm price, rainfall, risk at all levels and input prices. The situation of rice production in the state can be augmented only if policy prescriptions are launched by the government to make the farmers risk bearers. Yield can be improved by adopting better technology involving adequate, efficient, effective right type of inputs.

9

CROP DIVERSIFICATION IN ORISSA: AN ECONOMETRIC ANALYSIS

JAGANNATH LENKA

INTRODUCTION

Development being at the center stage in the policy agenda of almost all countries in the world has not only led to fast depletion of natural resources but also thrown up new challenges to the agrarian sector endangering the viability and sustainability of farm enterprises. There is a lurking fear that the fast changing global economic environment particularly after WTO might give a big jolt to the agricultural sector in the developing countries .The problem of these countries are further compounded by their inherent deficiencies in infrastructure both social and physical.

Diversification of agriculture in favour of commercial and high value crops is in fact recognised as a powerful strategy to counter the emerging challenges. As a risk management device crop diversification towards the crops in which a country has comparative advantage as compared to imports was traditionally assumed to be efficient in facing international competition. But now it has become inevitable as free trade in agriculture mandated under WTO has posed a constant threat to the developing countries in the form of high market induced risks and fall in global prices for farm products that have adversely affected traditional crops and families depending upon them.

The stiff competition in world market that the country's agricultural sector has to face in post globalisation period can be met with high degree of crop diversification only. The demand for food grains being highly inelastic, the crops with distinct qualities favourable to both domestic and international consumers should be encouraged.

With this backdrop the present paper makes a modest attempt to analyse the trends and determinants of agricultural diversification in the state of Orissa.The paper is organized as follows. The second section presents the post reform challenges. Data and Methodology are given in Section–III. Results and discussion are presented in Section-IV .The concluding section contains summary of the findings and policy options.

POST-REFORM CHALLENGES

A significant change in Indian Agriculture has taken place in post reform era due to liberalization and globalization through increasing export orientation and market access to others. To cater to these requirements the National Agricultural policy (NAP), 2001 emphasized the need for efficient use of soil, water, bio-resources, fertilizers etc. The NAP also visualized technological transformation of Agricultures so as to cultivate location specific and economically viable varieties of crops, use of biotechnology, strengthening the linkage between research and extension activities, provision of infrastructural facilities, and legalisation of tenancy to promote contract farming system.

In the context of the policy changes after economic reforms, the change in the production pattern in agricultural sector and its participation in world trade is noteworthy. Agricultural export hovered around 6 billion dollars with a declining share of cereals overtime. Though the production of food grain has increased from 176 MT from 1990-91 to 212MT in 2001-02 the area under food grain has shown a declining trend but there has not been much shift in the area under food grains to nonfood grains.

As compared to other states Orissa has not shown remarkable progress in agrarian development in post reform period. The State shows fluctuating and declining agricultural performance that can be attributed to natural calamities. Cereals still remain the dominant crops with little diversification of commercial crops confined to a few districts of Orissa.

Orissa is predominantly an agrarian state with 84% of its total population having agriculture as their mainstay of life. Agricultural labour accounts for 73% of the economically active labour force of the state. The cropping pattern in Orissa both in rainfed and irrigated situation is rice based that is a highly water consuming crop. The rice invites a low market price lowering the net return over cost. Orissa can divert 10 lakh hectares of land out of 45 lakh hectares presently under rice towards the cultivation of Cash crops without hampering the total supply of rice in the state. This diversification can raise farm income, check distress sale of the food grain and provide employment in farm sector when the crop is affected by natural calamities. The farmers definitely prefer that combination of crops from which they can derive maximum possible net revenue at least possible risk that will sustain their income on the long run.

DATA AND METHODOLOGY

The present study makes use of secondary data drawn from statistical Abstracts of Orissa, Economic Survey and Orissa Agricultural statistics. The analysis is based on the data at two points of time 1992-93 and 2002-03.

A change in the agricultural pattern is measured through the variation in the proportion of area under different categories of crops in the gross cropped area. For measurement of agricultural diversification, Herfindahl Index (HI) is used in the present study.

$$HI=\sum_{i=1}^{n} P_i 2$$

where Pi refers to the proportion of area under ith category of crops in the Gross Cropped Area (GCA). It takes the value between 0 and 1, when HI=0,it indicates perfect diversification whereas HI=1 presents perfect concentration.

The determinants of crop diversification is analysed using the linear regression model

$$Y = \alpha + \beta_1 X + \beta_2 X_2 + \beta_3 X_3 + \beta_4 X_4 + \beta_5 X_5 + \beta_6 X_6 + U$$

Where

Y = Herfindahl Index of crop diversification.
X_1 = Fertiliser consumption per hectare of GCA
X_2 = Proportion of irrigated area in the GCA
X_3 = Rainfall in millimeters
X_4 = GCA per tractor/trailor.
X_5 = Crop loan
X_6 = GCA per Regional cooperative marketing Societies.
U = Stochastic disturbance

In order to study the impact of these factors as crop diversification, a cross section regression model is estimated taking HI of crop diversification as dependent variable. The positive sign of the coefficient indicates declining diversification and negative sign indicates the increasing diversification.

RESULTS AND DISCUSSION

Cropping Pattern

A number of crops are grown in different districts of the state depending upon the local agro climatic conditions. For simplicity of the analysis these crops have been divided into nine broad categories such as Cereals, Pulses, Oilseeds, Vegetables, Sugarcane, Spices, Tobacco, Fibres and Fruits. In order to understand the changes in the cropping pattern that have been taking place at the state level from 1992-93 to 2002-03, the proportions of the area under crops grown to gross cropped area have been estimated and presented in Table 1.

It is observed that the major crops grown in the state belong to the Cereals. It covered an area of 64 per cent in the year 1992-93 and increased to 71.79 per cent in the year 2002-03 registering an increase of 12.16 per cent during the study period. The same trend is also observed in almost all the districts of the state.

Table 1
Area under Different Crops as Proportion of Gross Cropped Area

	Year		
Crops	*1993-94*	*2002-03*	*% Change*
Cereals	0.6400	0.7179	12.16
Pulses	0.1347	0.0996	-26.04
Oil Seeds	0.0693	0.0458	-33.92
Vegetables	0.1202	0.0975	-18.90
Sugar cane	0.0019	0.0024	25.97
Spices	0.0244	0.0243	-0.56
Tobacco	0.0014	0.0006	-54.59
Fibers	0.0080	0.0119	48.24
Fruits	0.0186	0.0328	76.34

The next major crops in the state are Pulses and Vegetables. Area under these crops as a proportion of GCA stand in the order of 13.47 per cent and 12.02 per cent respectively in the year 1993-94.It is found that area under Pulses as a proportion of GCA has declined by 26.04 per cent in the year 2002-03 while it has declined by 18.9 per cent in the case of vegetables.

In the case of Oilseeds this figure has also declined from 6.93 per cent in the year 1993-94 to 4.58 per cent in the year 2002-03.

Area under Sugarcane, Spices, Tobacco, Fibres and Fruits constitute only around 3 per cent of the GCA in the state. It is observed that the proportion of area under sugarcane, fibers and fruits have increased while that under spices and tobacco have declined during the period of study.

Though there has some changes in the cropped pattern in the state during the period 1992-93 to 2002-03,the position of the cereals has increased. The area under the commercial crops like Oilseeds and Vegetables, Spices and Tobacco has declined which is a matter of concern particularly after economic reforms. A clear shift in favour of the cultivation of Fruits, sugarcane and fibers is observed during this period.

Crop Diversification Indices

Herfindahl indices are estimated to show the crop diversification across the districts of Orissa.These estimates are shown in Table 2. A cursory glance at the HI values for the state as a whole (0.4647 for the year 1993-94 and 0.5456 for the year 2002-03) reveals the fact that there is crop concentration in stead of crop diversification during the study period. It is supported by the observation in the previous section that there has been increasing area under cereals.

Table 2
Herfundahl Index of Crop Diversity in Various Districts of Orissa

District	*1993-94*	*2002-03*	*% Change*
Angul	0.3715	0.4122	10.95
Balsore	0.5471	0.6821	24.67
Baragarh	0.2227	0.7366	230.68
Bhadrak	0.7299	0.7024	-3.76
Bolangir	0.4125	0.5358	29.89
Boudh	0.4754	0.5257	10.59
Cuttack	0.3628	0.4681	29.00
Deogarh	0.5286	0.4086	-22.69
Dhenkanal	0.4347	0.6611	52.07
Gajapati	0.4100	0.4031	-1.69
Ganjam	0.4304	0.5167	20.03
Jagatsinghpur	0.4357	0.4426	1.57
Jajpur	0.4665	0.5125	9.84
Jharsuguda	0.5440	0.6659	22.41
Kalahandi	0.4042	0.5789	43.23
Kendhamal	0.2649	0.3274	23.60
Kendrapara	0.5437	0.5148	-5.32
Keonjhar	0.5068	0.2618	-48.35
Khurda	0.5010	0.5270	5.19
Koraput	0.4060	0.4795	18.08
Malkangiri	0.4639	0.5742	23.78
Mayurbhanj	0.5268	0.6683	26.87
Nabarangpur	0.5402	0.6173	14.28
Nayagarh	0.4503	0.4997	10.97
Nuapara	0.4304	0.6030	40.08
Puri	0.5768	0.6560	13.72
Rayagada	0.3060	0.3303	7.95
Sambalpur	0.5533	0.6833	23.50
Sonpur	0.5853	0.7280	24.38
Sundargarh	0.5101	0.6222	21.97
Orissa	0.4647	0.5450	17.27

At the district level, crop diversification is found to be highest in Keonjhar followed by Deogarh, Kendrapada, Bhadrak and Gajapati.The rest of the districts are observed to have concentration of crops and it is highest in Baragarh followed by Dhenkanal. Thus it is very difficult to attribute any geographical and regional factors responsible for such diversification or concentration of crops across the districts of Orissa.

Factors Influencing Crop Diversification

Several factors influence the pace of agriculture diversification from staple food to high value commercial crops. Crop diversification is generally possible when the conditions required for various crops exists for growing varieties of crops. The required conditions include water resources, suitable soil, moisture and climatic conditions, use of seeds, fertilizers and pesticides, agricultural equipments, availability of finance, insurance provisions and facilities for marketing of the produce. The present study examined these factors of crop diversification and it is found that the most significant factors influencing crop diversification are fertilizer consumption, number of tractors and trailors, regional cooperative marketing societies, rainfall, irrigation and finance. The estimated value of the parameters of the regression for 1992-93 and 2002-03 are presented in Tables 3 and 4 respectively.

The model estimated for 1992-93 reveals that the factors retained in the model together explain 67.32 per cent of the variation in the crop diversification index.

Fertilizer consumption, proportion of irrigated area, GCA per tractors/trailor, crop loans, GCA per RCMS have positive impacts on the crop diversification. But the coefficient of fertilizer consumption and GCA per RCMS are not statistically significant. On the other hand rainfall has a negative impact on crop diversification and its coefficient is statistically significant.

In 2002-03, factors considered for explaining crop diversification could explain only 53.29 per cent of the variation in crop diversification index. It is observed that the factors that were exercising positive influence on crop diversification continues to do the same in the year 2003-04.While the coefficient of fertilizer is found to be statistically not significant, that of the other factors are statistically significant .The only factor retained in the model having negative impact on crop diversification is rainfall and its coefficient is not statistically significant.

Table 3
Estimated Value of the Parameters of Regression Model–1993-94

N = 30 — *Adjusted R^2 = 0.6732* — *Standard Error of Estimate = 0.053*

Variables	β	*Standard Error*	*t*
Constant	0.0358	0.1843	0.1942
Fertiliser consumption per hectare	-0.3295	0.5134	-0.6418
Proportion of irrigated area	-0.0026***	0.0014	-1.8571***
Rainfall	0.0002**	9.981E-05	2.1683**
Gross Cropped Area per tractor/ trailors	-0.3215*	0.1052	-3.0561*
Crop loans	-0.2124**	0.0829	-2.562**
Gross Cropped Area per RCMS	-0.0031	0.0056	-0.5536

* Significant at 1 per cent level
** Significant at 5 per cent level.
*** Significant at 10 per cent level.

Table 4
Estimated Value of the Parameters of Regression Model–2002-03

N = 30 — *Adjusted R^2 = 0.5329* — *Standard Error of Estimate =0.0291*

Variables	β	*Standard Error*	*t*
Constant	0.2119	0.0899	2.3575
Fertiliser consumption per hectare	-0.5124	0.4615	-1.1103
Proportion of irrigated Area	-0.0043**	0.0016	-2.6875**
Ranifall	0.0014	0.0123	0.1138
Gross Cropped Area per tractor/ trailor	-0.8751**	0.4128	-2.1199**
Crop Loans	-0.0027***	0.0015	-1.8000***
Gross Cropped Area per RCMS	-0.0021**	0.0008	-2.5547**

* Significant at 1 per cent level
** Significant at 5 per cent level.
*** Significant at 10 per cent level.

CONCLUSION

The main findings that emerged from the above discussion are as follows:

There is crop concentration rather than crop diversification in the state. The pattern of cropping has shifted in favour of cereals during the study period. Even after more than one and a half decades of economic reforms, commercial crops have not made any dent in the state except in some districts.

The factors that have significant influence on crop diversification such as fertilizer consumption, Proportion of irrigated area, GCA per tractor/trailor, crop loans and GCA per RCMS are found to be weak in the state for which there is crop concentration.

Specific suggestions that we want to make in the interest of the state are:

(i) Government should take immediate steps to facilitate and encourage crop diversification in the state.

(ii) NGOs and Panchayati Raj Institutions may be involved in creating awareness among the farmers regarding the importance of crop diversification.

(iii) Funds for Research and Development be earmarked so as to identify crops best fit for soil and water in different regions.

(iv) Finally, Foreign Direct Investment may be encouraged for development of agricultural infrastructure so as to facilitate agricultural diversification on the one hand and soil and water resource management on the other.

REFERENCES

Sekhar C.S.C. (2003),"Agricultural Trade Liberalization–Likely Publication for Rice Sector in India", *Indian Journal of Agricultural Economics (IJAE)*, Vol. 58, No. 1, Jan-March.

Saleth, Rmaria (1997)," Diversification Strategy for Small Farmers and Landless: Some Evidence from Tamil Nadu", *IJAE*, Vol. 52, No. 1, Jan-March.

Sharma H.R. (2005)," Agricultural Development and Crop Diversification in Hinachal Pradesh Understanding and Patterns, Process, Determinants and Lessons", *IJAE*, Vol. 60, No. 1, Jan-March.

Gopalppa, D. V. (1996), "Crop Diversification and Income Levels in Karimnagar Districts of Andhrapradesh", *IJAE*, Vol. 51, No. 3, July-Sept.

Vyas, V. S. (2000), *Changing Contours of Indian Agriculture.*

Dr. Utpal Kumar (2003), "Changing Cropping System in Theory and Practice: An Economic Insight into the Agrarian West Bengal", *IAJE*, Vol. 58, No. 1, Jan-March.

Chadha, G.K. (2003)," Indian Agriculture in the New Millenium: Human Resources to Technology Challenges", *IJAE*, Vol. 58, No. 1, Jan-March.

Kalamkar, S. S. & Pradhan, P. C. (2003), " Diversification of Agriculture in Orissa: Progress and Possibilities", *Agrarian Transformation in Orissa: Issues and Evidence*, Vol. I.

Tripathy, P. K., Bhoi, T. Panda Pravati (2004)," Agricultural Sector in Orissa: Growth Profile and Crop Pattern Changes", *Agrarian Transfornmation in Orissa: Issues and Evidence*,Vol. II.

Ratnam, C. & Mohan, Geetha (2005), " Crop Diversification in Andhra Pradesh: Trends and Determinants", *Souvenir*, 88 Annual Conference of the Indian Economic Association, December.

Jalal, R. S. (2005), *Agreement under WTO and Need for Agricultural Reforms.*

10

FERTILISER USE AND ITS CONSTRAINTS IN PADDY CULTIVATION: A STUDY IN MADURAI DISTRICT

S. Natchathira Jothi

INTRODUCTION

India is the second largest producer of rice in the World. But its National average yield is low compared to other advanced rice producing countries. The low yields are due to many factors such as high temperature, heavy or low rainfall and low light intensity during monsoon season etc. In order to obtain high yields, it becomes necessary to determine the optimum agronomic practices suited for different varieties. Among these practices, proper spacing, optimum fertilization, judicious water management, timely plant protection and water control are important. India's soil though varied and rich in plant nutrients is being depleted by intensive cultivation of crops. Even though fertilizer is one of the costly inputs, it is used to maintain fertility and also to increase production.

Hence a study on the influence of major factors on yield and the consumption of fertilizer in rice crop is very important to increase the yield of paddy. In this study an attempt is made to analyse the fertilizer use and its constraints in paddy cultivation in Madurai District.

OBJECTIVES

The main objectives of the study are:

(a) to study the cost and return structure and to estimate the fertilizer cost to total cost of I.R. 20 variety of rice in Thiruppalai village during the period October 2006 to March 2007;

(b) to identify the factors influencing yield and the consumption of fertilizer in rice crop and

(c) to examine the relationship between farm size and fertilizer use.

Hypotheses

The hypothese of this study are as follows.

(a) The cost of inputs like fertilizer, irrigation, seed and pest and farm size have positive impact on yield;

(b) Fertiliser use varies directly with the size of the farm.

Scope of the Study

The present study aims at analyzing the fertilizer use and its constraints in paddy cultivation at Madurai District. It covers a period of six months period- October 2006 to March 2007. Inputs such as Human Labour, Bullock Labour, Seeds, Organic Manure, Inorganic Fertiliser and Pesticides and output Yield are considered both in terms of quantity as well as money. The study concentrates on the aspects like Input-Output structure, cost-Return structure, Determinants of yield, share of fertilizer cost in total cost and constraints in fertilizer use.

Limitations

Only Samba season has been considered for this study since medium duration varieties like I.R. 20, Ponni, culture Ponni etc are being grown in this season. Since most of the farmers grow I.R. 20 variety in the study area (Thiruppallai village), this variety alone has been taken into account.

DESCRIPTION OF THE STUDY AREA

Agricultural sector occupies a predominant position in the economy of Madurai District. It is the second most thickly populated district in Tamilnadu. There are two seasons with regard to paddy cultivation in Madurai district. One is called Kuruvai (Kharif) or Paddy first crop and this season is commenced from June to September. The season is called Samba (Rabi) or Paddy second crop and it is commenced from October to February. Medium duration varieties like I.R.20, Ponni are being grown in Samba season. IR20 was developed by International Rice Research Institute(IRRI) Manila, Phillippines. It is cross between IR 202-24-3 and TKM 6-18. This was introduced in Tamilnadu in 1969.

The study area-Thiruppalai Village-comes under Madurai North Taluk, Madurai West Panchayat Block and Madurai district. This taluk is chosen because of two reasons. The major source of irrigation is canal. It is a relatively more developed taluk, modern techniques of agricultural production have entered on a large scale. This village is the only village, which is highly progressive in nature and main crops are paddy and sugarcane. The major source of irrigation is canal belongs that to Periyar-Vaigai irrigation project.

METHODOLOGY

It is an empirical study; both primary and secondary data were collected. A stratified random sampling technique was used to collect primary data from 60 farmers growing I.R 20 variety of rice. Out of 400 farmers who are the regular growers of paddy for two seasons, 30 small farmers with land holding less than or equal to 5 acres and 30 large farmers with land holding greater than 5 acres were considered. The study covers the Samba season of October 2006 to March 2007.

Information regarding the pattern of cultivation, types and quality of fertilizer applied and the extent of application, cost of cultivation, constraints felt by the farmers in the use of fertilizer, the availability of fertilizer and their preference towards the types of fertilizer were elucidated from the farmers through personal interview method by administering a well structured questionnaire. The major sources of secondary data were Fertilizer Statistics published by Fertiliser Association of India, New Delhi, pamphlets collected from the Thiruppalai village office and the Information book of Madurai West Panchayat union and published and unpublished sources like books, magazine, reports, journals etc.

Simple Linear Regression Model has been used to analyse the relation between fertilizer use and farm size. Multiple Linear Regression Model is used to examine the relation between the Inputs and output. To know the opinion of the farmers regarding the constraints in fertilizer use in paddy cultivation, Garret's Ranking Technique has been employed

RESULTS AND DISCUSSION

Input-Output Structure of I.R. 20 Variety of Rice

Human Labour, Bullock Labour, Seeds, Organic Manure, Inorganic Fertiliser and Pesticides are the major inputs used by the farmers in the study area. Table 1 portrays the input output structure per acre of I.R.20 variety of rice. It explains the extent to which cultivation of I.R.20 variety of rice is beneficial to the farmers.

Table 1
Input-Output Structure of I.R.20 Variety of Rice Per Acre

Particulars	*Quantity*
Human Labour	112.79 Man Days
Bullock Labour	10.175 Pairs
Seeds	30.5 Kg.
Organic Chemical	459.79 Kg.
Inorganic Fertiliser	160.82 Kg.
Pesticides	12455.50 Ml.
Yield	2012.39 Kg.

By using the inputs such as Human Labour, Bullock Labour, Seeds, Organic Manure, Inorganic Fertiliser and Pesticides, the farmers are able to produce nearly 2012.39 Kg of Paddy per acre.

Cost and Revenue Structure

First thing to be noted is that Human Labour consists of permanent and temporary hired Labour. There is no involvement of family Labour. Secondly since this village comes under Periyar-Vaigai irrigation project system, major irrigation system found here is canal irrigation system. This study is limited only to the Samba season, which includes six months. Hence, the rent per annum was divided by two to obtain the rent for six months.

Table 2
Cost and Return Structure of I.R. 20 Variety of Rice Per Acre

Components	*Valuesin Rs.*	*Percentage of Operational Cost*
Human Labour	816.45 (399.53)	38
Bullock Labour	129.65 (47.36)	6
Organic Manure	208.75 (63.48)	10
Inorganic Fertiliser	424.95 (127.59)	20
Pesticides	209.56 (83.57)	10
Seeds	120	5
Irrigation	(24.10)49.45	2
Interest on Working Capital	(9.71)195.46	9
Total Cost 'A'	2159.27	100
Rent for Six months	1752.50	
Total cost 'C'	3911.7	
Yield	5046.50	
Net Income over Cost 'A'	2887.23	
Net Income over Cost 'C'	1134.73	

From Table 2, it is understood that human labour cost plays a vital role. It consisted 38 per cent in the Total Cost 'A'. Irrigation cost is only Rs.49.45 (i.e 2 per cent). Total Cost 'A' (Operational Cost) plus rent for 6 months gives Total Cost 'C'. Net income over Cost 'A' is obtained by subtracting Total Cost 'A' from yield in Rupees. Similarly Net-Income over Cost 'C' is the difference between Yield in Rs. and Total Cost 'C'. Human Labour Cost and Fertiliser Cost constituted the major portion (i.e 68 per cent) in the Total Cost 'A'. Fertiliser cost alone constitutes nearly 30 per cent of the Total Cost. Cost incurred for irrigation is very low (i.e only 2 per cent).

Determinants of Yield

The factors such as Labour cost, Fertiliser cost, Irrigation cost, Seed and Pesticides costs were combined together in the model as their individual shares worked out to be very low. Dummy variable was introduced to assess the impact of differences in the farm size. In this study, a sample of 60 farmers is taken into account. Here yield per acre is considered to be the function of labour cost per acre, fertilizer cost, irrigation cost etc. The variation in farm size is represented by Dummy variable. D = 0, if the farm size is small, D = 1 if the farm size is large.

Table 3
Determinants of Yield

Variables	*Parameter Estimates*
Intercept	4269.82
a_0	2.447*
X_1	(1.73) 1.245*
X_2	(1.89) 0.178**
X_3	(1.47) 1.397**
X_4	(1.41) 8.467**
b	(1.57)
R^2	0.698
F	24.940
No. of Observations	60

Figures in parentheses represent t-value.
* - Indicates significant at 5 per cent level.
** - Indicates significant at 10 per cent.

The MLR Model fitted for the identification of determinants of rice yield was of the following form: $Y = a_0 + a_1 X_1 + a_2 X_2 + a_3 X_3 + a_4 X_4 + b D + U$, Where Y – Yield in Rs./acre, X_1-Labour Cost, X_2 –Fertiliser Cost X_3 – Irrigation Cost, X_4 – Seeds and Pest cost, D –Dummy variable for variation in farm size. The disturbance term U is normally distributed with 0 mean and variance s^2. This equation was estimated with the help of the principle of Least Squares. Table 3 gives the results of the estimated regression.

The analysis revealed that fertilizer to be the most important determinants of rice yield. It was followed by labour.

All the independent variables have theoretically expected sign. Totally the independent variables account for nearly 69.7 per cent variations in the yield. Labour and fertilizer have emerged statistically significant at 5 per cent. All the independent variables have a positive impact on yield. When independent variables such as labour and fertilizer increased by one per cent, yield could be increased by 2.45 per cent and 1.25 per cent respectively.

Managerial efficiency has been measured using Dummy variable. One being allotted to large farmers having more than or equal to 5 acres and zero to small farmers owning less than 5 acres. The signs and value of the coefficient for the dummy variable implies that large farmers apply more fertilizer than the small farmers. This may be due to the following reasons. Since large farmers are normally rich, they are able to get all the inputs at right time. So they can adopt intensive cultivation by which they will be able to get more net return. Small farmers at time are forced to sell the goods through intermediaries because of the pressing problem due immediate settlement of their debts. The surplus product could be kept for the future use by the large farmers since they have modern and good storage facilities.

Since the F-value is significant at one per cent, the model could be said to be statistically significant. The positive impact of labour, fertilizer and other determinants on yield indicates that land productivity may be increased by their additional use. In the new technological era in agriculture, the emphasis is more on proper fertilizer use by the farmers. Farmers may be made educated on the significance of the input use in raising yield.

Farm size and Input-Output Relationship

Farm size was categorized as 1-3 acres, 3-5 acres, 5-7 acres and above 7 acres. On the basis of this, number of farmers, average input used and yield were found out and furnished in Table 4.

Table 4 highlights cost structure (input) and average yield (output) per acre. It is easy to understand that there is positive relation between farm size and output. It means that the farmers cultivating larger farms are financially sound naturally and they possess the capacity and intensity to spend more to get maximum yield. It is important to note that in farm size 1-3 acres and 3-5 acres, when the Average Cost spend per acre increased from Rs. 1874.68 to Rs. 1937.65, there is corresponding increase in yield from Rs. 4872.51 to Rs. 4953.92, while the cost is Rs. 1987.69 and 2062.86 in the case of farm size 5-7 acres and more than 7 acres respectively.

Table 4
Farm Size and Input-Output Relationship (in Rs.)

Farm size (acres)	*No. of Farmers*	*Average Yield*	*Average Cost* Labour	Fertiliser	Seeds & Pests	A. Total Cost
1-3	18	4872.51	907.5	595.12	324.56	1874.68
3-5	12	4953.92	941.94	619.75	327.75	1937.65
5-7	18	5152.57	966.28	640.30	331.06	1987.69
Above 7	12	5241.11	977.88	695.63	336.63	2062.66

Still there is an increase in the yield from Rs. 5152.57 to Rs. 5241.11. This may be due to economies of scale. Labour and fertilizer costs are found out to be the two important costs, which have significant influence on yield. That is both inputs together contribute over 68 per cent to the Cost 'A'. There exists direct relationship between input and output.

Share of Total Fertiliser Cost in Total Cost

From the information given in Table 5, it is understood that there exists a positive relation between Farm size and total fertilizer cost. The share of fertilizer in total cost was 31.75 per cent in the case of small farmers and it is 33.72 per cent for farmers having more than 7 acres.

Table 5
Share of Fertiliser cost in Total Cost

Farm size (Acre)	*Total Cost Rs./Acre*	*Fertiliser cost Rs./acre*	*Share of Fertiliser cost in Total cost (%)*
1-3	1874.68	595.12	31.75
3-5	1937.65	619.75	31.98
5-7	1987.69	640.30	32.21
Above 7	2062.86	695.63	33.72

Hence, it is clear that more amounts are spent by farmers both in terms of money as well as proportion when the farm size increases.

To examine the relation between farm size and fertilizer use, SLR Model was fitted. Y = a + b X; Y – quantity of NPK in Kgs/acre, X – Farm size in acres and a – intercept. The analysis revealed that fertilizer use varies directly with the size of the farm. For one unit increase in the variable X, namely farm size, the quantity of NPK consumed increases by 1.17 unit. This independent variable has theoretically expected sign and has emerged statistically significant at 5 per cent level. This variable accounts for nearly 62.9 per cent variations in the yield. This model could be said to be statistically significant since F-value is significant at 1 per cent.

Constraints in Fertiliser Use

To know the opinion of the farmers regarding the constraints on fertilizer use the famous qualitative technique namely Garret's Ranking Technique was used.

Table 6
Constraints in Fertiliser Use

Constraints	*Total Score*	*Mean Score*	*Rank*
Uncertainty in the availability of irrigation water	4710	78.50	I
Non availability of credit at appropriate time	3722	62.03	II
Non-availability of fertilizer in time	3655	60.97	III
High prices of fertiliser	2880	48.00	IV
Sources far away from the village	2670	44.50	V
Income not commensurate with the level of application	1745	29.08	VI
Farm size is too small to apply fertiliser	1615	26.92	VII

First step is to calculate the per cent for each problem (Constraint) ranked. This was done by using the formula: Per cent = [100 (R_{ij} – 0.5)] / R_j where R_{ij} – Rank given to particular problem and R_j – total number of problems ranked.

Thus Rank I shows that the uncertainty in the availability of irrigation is the most urgently felt. Second important factor is the non-availability of credit at appropriate time.

SUGGESTIONS AND CONCLUSION

For the efficient use of fertilizer the following suggestions could be taken into considerations. First and foremost constraint on fertilizer use is the uncertainty in the availability of irrigation water. There are many wells and tanks but almost dry. So it is imperative on the part of the government to take necessary steps to provide proper irrigation facilities. Measures should be taken to make the agriculturists to utilize the services rendered by the agriculture department fully, especially in the areas like soil test, proper application of fertilizer etc. Agriculture department should come forward to conduct meeting and demonstrations frequently in the village. More branches may be opened by nationalize banks and liberal loan schemes may also be evolved so as to cater the credit needs of the farmers. Steps may be taken to supply more fertilizer through Co-operative Society and to encourage businessmen to start retail outlets in the village. Seeds, Fertiliser and pesticides could be given to the small farmers on subsidized prices since they are financially backward and in need of financial help.

Crop insurance scheme must be made easy and available to all crops. Transport facilities must be improved and the basic amenities like water, electricity must be extended to the village. Many more T.V.Programms may be telecast to make the farmers to accept new techniques and to know the facilities provided by the government. Installation of educational institutions and proper implementation of adult education programme may help to minimize illiteracy. This would help them to avoid middlemen in marketing their produce and guide them to use government warehouses and co-operative marketing federations to market their products.

Professor Cairncross has said, "No country can count itself developed, in which education in the way of industrial civilization has not taken place. Peasants have to be brought within the monetary economy and not left to pursue subsistence farming".

REFERENCE

Arora R. C., (2003), Development of Agriculture and Allied Sectors, S. Chand & Coy Ltd., New Delhi.

Brahmananda P. R., (1982), Productivity in the Indian Economy-Rising Inputs and Falling Outputs, Himalaya Publishing House.

Anandaraj D. Sunitha, (2004), Farm Size and Fertiliser use in Andhra Pradesh, *Indian Journal of Agricultural Economics,* Vol. 41, No. 4, Oct-Dec .

David Hopps W. (2003), The Economics of Fertiliser use-A Case Study on Rice Production, *Indian Journal of Agricultural Economics,* Vol. 17, No. 4, Jan.-March.

Singh B. K. and J. S. Sharma (2001), Economically Optimal Fertiliser Requirements for Wheat and Paddy Crops in Different Regions of U.P, *Indian Journal of Agricultural Economics,* Vol. 39, No. 4, Oct-Dec.

Shenoi P.V., (1998), Agricultural Development in India, Vikas Publishing House Pte., Ltd.

Antle M. (1989), The New Economics of Agriculture, *American Journal of Agricultural Economics,* Vol. 81, No. 5.

Manmohan Singh (1997), The Unfinished Business of Nation-Building, *The Hindu,* 15.

Agricultural Statistics at a Glance, Directorate of Economics and Statistics, Department of Agriculture and Cooperation, Ministry of Agriculture, Government of India. *http://agricoop.nic.in/about/htm.*

11

TECHNOLOGICAL CHANGE AND FACTOR SUBSTITUTION IN INDIAN AGRICULTURE

Harish Mani, G. Bhalachandran,
C. H. Shah & V. Pandit

INTRODUCTION

The interest in factor substitution can be traced to two major sources. First, T.W. Schultz's seminal work (Schultz, 1964) kindled the interest in technological change in agriculture and the concomitant interest in the structure of production and the structure of input-combinations. Schultz's famous statement characterizing traditional agriculture as 'efficient but poor...', pointed to the factor substitution especially between land on the one hand and other factors on the other. Schultz finally ended with the technological change that facilitated substitution of land with new inputs like new seeds and fertilizers. He also recommended investment in human capital, which was well received. Second, the dual economy model developed by Arthur Lewis assumed zero marginal product of labor in agriculture as Schultz did. By implication, it suggested low or no substitutability among land, labor, and capital in agriculture.

The remedies to the situation of limit to the further economic growth in general and of agriculture in particular suggested by the two leading professionals differed. Schultz emphasized transformation of agriculture through technological change for releasing the production constraint. In his scheme, the key role was to be played by new and improved inputs. If along with the new opportunities for agricultural production to expand, the industrial sector of the economy also expands, the limit to the growth of the economy could be loosened and the economy as a whole may begin expanding. The dual economy model concentrated on moving labor out of agriculture while the food, the main requirement for labor to sustain would be reallocated.

In the above construct, the role of population has not been explicitly mentioned. In his recent work Nerlove (1996), has brought in population as an endogenous factor along with environment. The terms of trade along with other variables finally influence the release of choking off of the growth of the economy. Terms of trade in turn are affected by the growth of agriculture relative to the expansion of population of the nation. In the end, the rate of growth of population and the technological change in agriculture determine the development possibilities for the economy as a whole. Nerlove's model suggests that the rate of technological change in agriculture has to be higher than the rate of growth of population for smooth and continuous development of the economy.

One special feature of agriculture which needs to be emphasized is that as a result of its limited supply, land has a dominant role in agriculture. With complementarity between labor and capital, industries can continue to expand if capital can be poured into the sector. Since land has a physical limit beyond which its supply cannot be increased, the other inputs can be combined in increasing amounts with limited agricultural land only if the substitution of land by other factors individually or together is made possible. There are serious limits to such substitution if the technology used for combination of factors and their contents are unchanged. Here in comes the role of embodied technological change in agriculture. New factors like fertilizers, and new seeds such as HYV are brought in with the new technology embodied in them. Therefore, agricultural production could expand only with continuous change in technology or what has now come to be known as 'Evergreen Revolution'.

The unending debate relevant to technological change in agriculture pertains to the issue of technological change being induced or autonomous. According to one view, population is the main actor in the issue of the induced technological change. The behavior of technological change as mentioned above, has led to the newly generated interest in research regarding the behavior of factor substitution in agriculture. Our attempt in the exercise undertaken by us is to delve into the changes in the factor substitutability over time. The data used pertain to the time series for the nation as a whole.

THE INDIAN EXPERIENCE

In the immediate post-World War II period, India experienced an unprecedented surge in population. After a long period, there are now some indications of a decline in the rate of growth of population. It may be worth observing the fact that during the period of population explosion, the food situation deteriorated steeply. The situation was so critical that it was described as a critical like 'ship to mouth' situation. The international agencies had reservations whether India could ever be helped to come out of the drought like situation. It was during this critical period that the search for ways to increase production at home became intensive. With a sense of risk, the HYV seeds of wheat were imported. The enterprising decision by the government paid off and the new move which came to be known as Green Revolution took off. The move became widespread and efforts increased its coverage to a wide variety of crops. It may be mentioned that the technique imported initially was known to the west since the early decades of the last century. The critical situation regarding the food supply led to its implanting on Indian soil. Can this be considered as an induced technological change, may be left for others to investigate appropriately.

It is important to note at this stage that even though the agricultural sector made an impressive progress in the recent decades, we are faced with a core problem of sectoral imbalance. Share of agriculture in GDP has declined rather rapidly from 58 per cent of GDP in 1950-51, to 20 per cent in 2005-06. However, the share of agriculture in the total work force has gone down from 76 per cent in 1961 to about 60 per cent in 1999-2000. Thus we see that the share of agriculture in the work force declined but at an agonizingly slow pace leading to the current imbalance. To understand the role of agriculture in the development of the economy, one has to have the total view of the growth strategy as a whole. This would require full-fledged modeling of the economy. One such exercise was done recently (Krishnamurty *et el* 2004). Authors observe that, '...the crowding–in of private investment in both agriculture and infrastructure, coupled with moderate initial increase and eventual decrease in price, led to real GDP increase through supply as well as demand.' The production function exercise we have done in this paper however, confines to the production relation of output and inputs based on which supply would respond to changes in prices. Thus, the objective of our exercise is limited but relevant.

It is our view that transformation of agriculture would be achieved through the technological change embodied in inputs. It would therefore be interesting to observe whether the technological change would or would not affect the relation of land with other inputs. Our investigation is confined mainly to two aspects. Firstly, the net technological change is identified. Then we turn to its impact on factor-factor relations of substitutability and complementarity. It is expected that the technological change would affect the factor-factor relationship, as the new technology helps substituting land, the supply of which is severely limited. The bundle of new inputs includes new form of capital along with the old form of capital. Bullocks continue to play their role as plough animals, but the introduction of tractors will change the composition of capital. Composition of capital input is expected to change also due to mechanization of irrigation, as a result of the introduction of oil engines and electric motors.

During the period under investigation, the use of power tillers was also introduced. It would be interesting to observe how the changing composition would affect the relation of capital with land and other inputs. To what extent the new technology would influence production will change over a period of time, depending on the degree to which the new technology is diffused over the area and crops covered. This may be referred to as the 'Spread Effect'. This changing impact of the new technology on production over a given time period will have an impact on the factor-factor relationships over the time period concerned. As we shall see the technological change may lead to increased marginal productivity of an input especially in the case of embodied technological change. The factor-factor relationship between land and other inputs is expected to be one of substitutability since technological change, be it embodied, neutral, or disembodied, helps in substituting land which is limited in supply, with other inputs such as HYVs, chemical fertilizers, irrigation etc.

MODEL AND DATA

Initially we have fitted a production function to explain the relationship between inputs and output. The form of the function considered is as follows:

$$Log(Y_t) = a_0 + a_1 Log(L_t) + a_2 Log(K_t) + a_3\{Log(L_t)*Log(K_t)\} + a_4 Rainfall$$

Where Y=Output, L=Land, K=Capital, and the subscript't' stands for time period.

Other variables such as, labor, irrigation and fertilizers were also considered, but due to inherent problems in the nature of these variables, they had to be excluded from the production function. One problem with these variables was that they had a very high correlation with capital, making it difficult to include them in the production function due to multicollinearity problem. In the estimation of the production function, a trend variable has also been introduced. This is mainly to account for Neutral technological change and also the effect of the inputs not explicitly considered in the production function for various reasons. The impact of embodied .technological change in capital has not been explicitly modeled in the production function but has been taken into account separately as will be seen below.

Following this, the marginal products of the two inputs land and capital are estimated. Increasing marginal products of land and capital would be indicative of technical progress. With respect to capital in particular, increasing marginal product would be indicative of an embodied technological change in capital. This is due to the following reason. Increased amounts of capital applied to a fixed amount of land leads to diminishing marginal productivity of capital. If on the other hand there is increasing marginal productivity of capital in the above scenario, it is indicative of an embodied technological change in capital, whose impact is more than the diminishing marginal productivity effect of increased capital such that the net effect on the marginal productivity of capital is positive. In the Indian context, the net sown area has more or less remained constant since about 1970 onwards. Given this fact, it would be interesting to observe the behaviour of marginal product of capital since 1970 onwards to identify any possibility of embodied technological change in capital.

Technological change also affects the relationship among the factors of production in terms of substitutability or complementarity. In a traditional system, the relationship among land and other factors of production is expected to be complementary. It is a purely land-based agriculture wherein additional amounts of land are needed to be brought under cultivation in order to increase output. But with technological change, more 'better quality' and nontraditional inputs can be applied to a fixed amount of land and thereby, increased levels of output can be generated. In this scenario of technological change, the relationship between land and other inputs is expected to be one of substitutability.

The time period of the study is 1950-51 to 1999-00. The variables considered in the final analysis study are output, land, and capital. Output is measured as real GDP in the agricultural (without allied activities) sector measured at 1993-94 prices. The unit of measurement of output is 'Rs Crore'.

Value of aggregate real net fixed capital stock calculated at 1993-94 prices is used as a measure for capital. The unit of measurement is 'Rs. thousand'. Due to non availability of necessary data, this composite variable had to be used as a measure for capital instead of using individual items like tractors, plough animals, oil-engines, electric motors and traditional and non-traditional implements. The latter approach, i.e., consideration of individual input contribution, seems to be more logical than taking a composite variable to measure capital.

Total net sown area has been taken as a measure for land. The unit of measurement is 'million hectares'. It is agreed that the area under current fallow justly belongs to land as an input, for, net sown area and current fallow together are a part of cultivation system. At ground level, the area under current fallow is influenced by widely varying definitions used by different states. The definition varies from two years to eight years period of non-cultivation of a piece of land beyond which, the same will be reclassified as cultivable waste land. The net sown area excludes intensity of cropping in terms of two or multiple crops. Alternatively, one can use gross cropped area as an input of land. But the area under annual crop will be considered as one-time cropped area. There is thus no completely satisfactory definition of land as an input. Intensity of use and quality of land are difficult to include in the account of land input with the available data, though, both are ideally part of land use in relation to the production of crops.

Likewise, total net irrigated area is taken as a measure for irrigation. The unit of measurement is 'million hectares'. Irrigation is important especially in monsoon regions. In India, for most parts, the rain is limited to four months and bulk of the rain is received in total sixty days. Irrigation helps to control water required for growth of crops.

Labor is yet another complex variable. Use of mere workforce for labor would be too simplistic. The ideal for labor input would be the hours of work, together with the quality of labor. Distinction can even be made between labor put in by owners, by tenants, and by wage laborers. We had to use a compromised variable due to limitation of data. Our data includes the effect of composition of male and female labor, level of education, and age of participants. This refinement adds to the quality of data but is not sufficient to overcome the original weakness. Since data are available once in every ten years, for intervening years, data were interpolated by Siva Subramonian. Thus we had five original readings and the rest were supplied. Variations from year to year are not real and the series is a smooth transition from the beginning to the end of the period.

The variable rainfall considered in the production function is the All-India monsoon rainfall index calculated by the Indian Institute of Tropical Meteorology, Pune.

Sources of data are mainly three; Data on land, irrigation and fertilizers are taken from the RBI's Handbook of Statistics on the Indian Economy. Data on output is taken from the National Accounts Statistics of India, and finally, data on labor and capital is taken from Sivasubramonian's (2005) stellar work, "Sources of Economic Growth In India - 1950-51 to 1999-2000". The Government publications have authentic data. S.Sivasubramonian's work contains data carefully selected and professionally processed for reflecting reality as closely as possible. To make the series continuous, he had to pare the data selected from different sources. Where data had gaps, he used interpolation as in the case of labor. As we have just observed, the original observations for labor are very few, and made up data were for much larger number of years. As a whole, where data required were not available, he compromised by using the next preferred proxy. All these procedures are acceptable at the professional level. Economic concepts are sharp but data belong to the real world and do not match the theoretical validity. This is not uncommon. But the short-comings take their toll in the econometric analysis.

A word of caution regarding land data is in order at this stage. Series begins with 1950-51. Land records were in the transition stage during fifties. In parts of the country governed by erstwhile alien rulers, land was cadastrally surveyed. The data were fairly accurate. In parts not governed by British, the accuracy varied very widely at the time of independence in 1947. Efforts were in process through out the following decade to make them as accurate as possible. At the fringe, the data regarding forest and cultivated area were imprecise, since the dividing lines between forests and cultivated area were not well defined. Besides, the cultivated area which covered net sown area was corrupted by the Government intervention to grapple with the post-war situation of food scarcity. Especially the policy of compulsory levy and price control for food grains led to under-reporting of area under crops. As the controls were relaxed, the land under-reported earlier came back on records. This fact would affect the trend in agricultural production. There is no easy way to deal with this weakness of the vital data regarding land under cultivation. We have to contend with the data available.

One more factor that would probably affect the exercise in some measure may be mentioned. The new technology spread over the states over a period of time. Initially, well-endowed states benefited from the new technology. Other states followed as the new technology with appropriate adjustment became available. The resource endowments varied widely over states and so, the benefit to the states varied over space and hence over time too. The effect of technological change over states on the trend may be marked. This may be called a 'spread effect'.

TRENDS IN OUTPUT AND INPUTS

In order to get a feel of the changing relations between inputs and output, a graph of the index numbers of inputs and output is given below (Base; 1950-51=100).

We find that the pattern of growth of output resembled the growth of inputs taken individually. Capital and land were the only exceptions; Land showed an increase as an input in fifties. Thereafter land was almost stationary. Capital outpaced output for a few years in eighties. Perhaps the improved performance of output in the preceding years encouraged expectation for the subsequent years which might have led to speedy investment. Soon in the following years, capital investment slowed down. The graph suggests that the growth in output once again outpaced the growth in capital.

It is interesting to note that irrigation continued to increase steadily through out the period. But the pace had slowed during eighties. The net effect of slow down in irrigation during

Index Nembers of Output and the Various Inputs

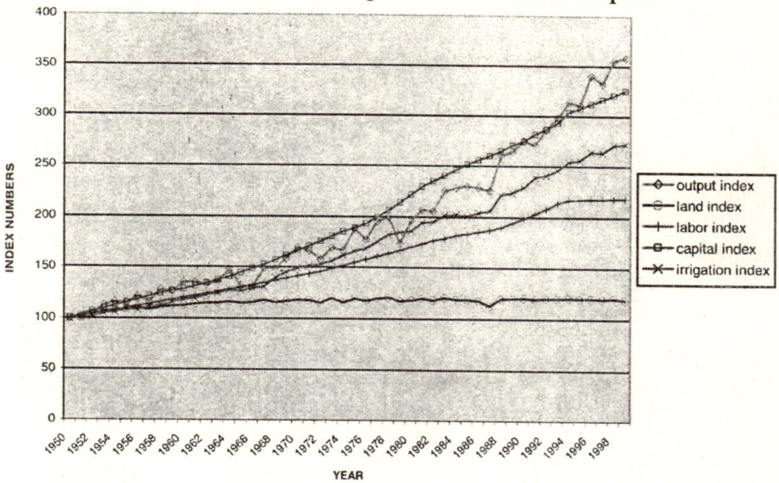

eighties could explain the outpacing of production by capital during the eighties. During the next decade, irrigation picked up pace and so also output.

Labor increased at a slower pace during the later years. This happened despite the fact that the quality of labor reflected in terms of education improved. One explanation could be that with rising incomes of important sections of the farming community, there was withdrawal of labor from the workforce, especially so in the case of young population of school going age and some of those of age above 60. With incomes rising, women too preferred to withdraw from the hard labor involved in agriculture, and perhaps preferred to focus on well being of family.

On the whole, the output grew at a steady pace for most part and, with an exception of eighties, the pace of output growth was higher than that of all inputs taken individually. This would indicate an implied benefit of technological change represented by Green Revolution.

EMPIRICAL RESULTS

Before the estimation of the production function, we carried out the stationarity tests on different variables using the Augmented

Dickey-Fuller test. The results of the test are given in the table below.

Table 1
Stationarity Tests Results

Variable	*Level*	*1st Difference*	*2nd Difference*	*Inference*
Log(Y)	-4.480505			Stationary- I(0)
Log(L)	-3.499071	-11.78439		Nonstationary- I(1)
Log(K)	2.131543	-3.708232	-10.46733	Nonstationary- I(2)
Log(L)*Log(K)	-2.738894	-11.81226		Nonstationary- I(1)

Using different variables in their stationary forms, we estimate the following production function estimated by OLS.

$$LOG(Y_t) = \underset{(134.69)}{10.96236} + \underset{(2.57)}{7.8571}\, D(\log\{L_t\}) + \underset{(1.62)}{1.4527}\, D(D(\log\{K_t\}))$$

$$- \underset{(-2.49)}{0.6140}\, D([\log\{L_t\}] * [\log\{K_t\}]) + \underset{(3.69)}{0.0724}\, DUM1 + \underset{(38.96)}{0.0262}\, TREND$$

$$+ \underset{(3.155)}{0.00026}\, RAINFALL + \underset{(3.03)}{0.4276}\, AR(1)$$

$$R^2 = 0.992$$

$$\bar{R}^2 = 0.991$$

$$D.W = 2.35$$

As noted from the above production function, all the inputs have a statistically significant relationship with output. Though rainfall is one of the explanatory variables in the estimated equation, it does not fully capture the outliers associated with drought years. For this reason, we also include a dummy variable DUM1 which takes account of five drought years namely 1952-53, 1965-66, 1966-67, 1979-80 and 1987-88. All these specific drought years made a significant impact on output as indicated by the significant't' value of DUM1.

We have introduced trend as an independent variable. It measures the impact of factors not taken as explicit variables. Trend variable also takes account of neutral technological change during the period. The trend value is highly significant and positive. It implies that factors like labor, fertilizers, HYV seeds, and macro level variables like infrastructures, consisting of transport, irrigation projects, extension, marketing, together have made a significant impact The trend coefficient of 0.0262, as shown in the above production function, is indicative of increase in production at the rate of 2.62% p.a. over the period 1950-51 to 1999-00.

Capital includes implements, plough animals, and machinery such as tractors, oil engines, electric motors, and allied ones including power tillers and combines in some cases. The composition of capital would vary over years as substitution among them took place over time.

Role of land in agriculture is direct and significant. Even when it increases slowly or does not increase at all, it allows other inputs to make contribution through it. This role of land makes it an important production agent. This is reflected in the significant and positive coefficient of land. We have not included land improvement as an input either as an independent variable to reflect quality of land, or as a part of land as a variable.

Technological Change and Its Measurement

We have introduced trend, representing time (years under study) as a variable. It can be considered as to represent technology. We use the term in a broad sense. The factors not tacitly included in the function will have their effect included in the coefficient of trend. So also, the technology not embodied in capital and land will also be impacting the value of coefficient of trend.

While the neutral technological change will shift the entire production function and the shift will be parallel, the embodied technology will be represented by a change in the slope when production function shifts. In terms of the above analysis, while the parallel shift will be taken care of by coefficient of trend, the change in the curvature or marginal values at different points will be represented and measured in terms of changes in marginal values of an input. A word of caution is that this measure too will be an approximation. The change in the curvature is not directly measured. The values of the marginal products of the two inputs land and capital is examined below

Table 2
Marginal Products of Land and Capital

Variables	*1951-60*	*1961-70*	*1971-80*	*1981-90*	*1991-00*
Land (million hectares)	3112.365	3478.326	4214.243	5590.164	7529.079
Capital (Rs. 1000 at 1993-94 prices)	0.110	0.093	0.086	0.089	0.099

At the very outset, the figures of the marginal products for land and capital look quite contrasting to each other. This is due to the units of measurement of the two inputs. Since the unit of measurement of land is 'million hectares', the interpretation of marginal product of land would be 'change in output per million hectare increase in the net sown area'. On the other hand, since the units of measurement of capital is 'Rs thousands', the interpretation of marginal product of capital would be 'change in output per thousand rupees increase in capital stock'.

A glance through the above figures shows that the marginal product of land has been increasing throughout and the increase has been substantial. In the case of capital, the marginal product has declined up to the decade 1971-80 and since then has increased. This behaviour of marginal products needs an explanation. First, let us consider the variable capital.

As observed earlier, for cultivation, new forms of capital that replaced the old ones had an element of improved efficiency or an element of technological change. Bullocks were replaced by tractors. Tractors enable to do deep ploughing. Deep ploughing helps turn soil and bring the nutrient below the upper soil, closer to the surface. Besides, tractors help speed up ploughing and preparation of land for second crop where it is permitted by weather and soil conditions. Similarly, the traditional wells were replaced by tube wells. Tube wells helped to reach the water available in aquifer below the upper strata. The tube wells enabled increasing area under irrigation.

Additional area was irrigated also by putting up additional new tube wells.

The decline in the marginal product of capital during the decade 1961-70 was due to the prevalence of traditional technology wherein, an increase in the amount of capital at a rate which is more than the rate at which the area under cultivation increases, leads to a decline in the marginal productivity of capital. As seen in the table above, the marginal productivity of capital had further declined during 1971-80. The decade 1971-80 was the period when the 'Green Revolution' was just picking up in the Indian agriculture. It was mainly in terms of using high yielding varieties in the place of traditional varieties of seeds, and with regard to capital, the replacement of traditional forms of capital with superior types of capital, etc. Since this process was just picking up during this decade, the diminishing marginal productivity effect of increased capital dominated the 'technological change' effect, with the net result being the decline in the marginal productivity of capital. During this period, the conventional forms of capital such as traditional ploughs and bullocks and traditional wells still dominated As this process of substitution of traditional capital with superior forms of capital gathered momentum especially in the later part of the period under study, namely 1981-90 and 1991-00, the effect of the 'technological change' was more than the diminishing marginal productivity of capital, such that the net effect was an increase in the marginal productivity of capital. The increment in value of marginal product would suggest increased embodied (new) technology. The total input of capital also increased, while quantity of land with which capital combined remained nearly unchanged. If the quality of capital had not improved with technological change embodied in capital, the marginal products would have tended to decline.

The increase in marginal product of land when its quantity remained nearly unchanged can be explained in terms of theoretical formulation. According to the Ricardian principle of diminishing marginal product, value of marginal product of land would tend to rise, as the input with which it is combined increases, with the area under cultivation remaining unchanged. Value of marginal product of the input combined with land diminishes. The theory of rent was based on this principle. When quality of capital improves, it would imply as if there was further addition to capital. Such stupendous increase in capital would lead to equally stupendous increase in marginal product of land. Our results match this expectation.

We conclude at this point that, not only did Green Revolution bring about a net shift in the production function as represented by trend, but it also facilitated the use of improved capital and irrigation inputs, i.e., it brought about embodied technological change, the technology being embodied in capital (and in other inputs not included in the study like irrigation, fertilizers, seeds and pesticides). The benefit seems to have increased over time suggesting continuity of the benefits of Green Revolution. To the extent the benefit derived from Green Revolution increased due to the spread of the use of new and improved inputs from initial area to additional area, it may be better termed as "spread effect".

As the shift takes place from a 'traditional agriculture' regime to a 'modern agriculture' regime, the relationship between land and other inputs, capital in particular, is expected to change from one of complementarity to one of substitutability. In the next section, the Allen's partial elasticity of substitution is calculated between land and capital for the five decades spanning the period of our study.

FACTOR SUBSTITUTION AND THEIR IMPLICATIONS

The levels and changes in the marginal products of the factors of production depend among other factors, on the substitutability or complementarity among the factors of production and vice versa. The Allen's partial elasticities of substitution were worked out between the two factors, land and capital, for each of the five decades spanning the period 1951 to 2000. The results are laid out in the table given below.

Table 3
Allen's Partial Elasticity of Substitution between Land and Capital

Decade	AES_{LK}	*Relationship*
1951 -60	15.304	Substitution
1961-70	17.184	Substitution
1971-80	18.674	Substitution
1981-90	18.236	Substitution
1991-00	16.402	Substitution

The above results are interesting. The relationship between land and capital which is one of substitutability, remains the same all through the five decades spanning the period of study. Also the change in the values of AES_{LK} over the decades has been only marginal. The results can be explained as follows. The high substitutability between land and capital observed from the decade 1971-80 onwards is largely attributed to the Green Revolution wherein superior forms of inputs, with capital in particular, were applied on a nearly fixed amount of land in order to generate more output. The decline in the value of the AES_{LK} during the decade 1991-00 could be attributed to the decline in the public investment in agriculture during the 90s which in turn would have 'crowded out' private investment into agriculture.

As already indicated above, in a traditional agricultural setup, the relationship between land and other inputs is expected to be one of complementarity. In this light, it seems rather surprising to see a substitution relationship between land and capital during the period 1951-70. The substitution relationship observed during the decade 1951-60 could be attributed to the large scale investments made in irrigation during the first plan period, which added a lot of capital stock to agriculture. The substitution relationship observed during 1961-70 is unexplainable at this stage.

COUNTERFACTUAL SIMULATIONS

We have carried out a series of counterfactual simulation exercises under various alternative sets of assumptions and compared the same with the baseline solution. The basic model considered for the purpose of simulations is given as follows -

$$LOG(Y_t) = 10.96236 + 7.8571\,D(\log\{L_t\}) + 1.4527\,D(D(\log\{K_t\}))$$
$$-0.6140\,D([\log\{L_t\}] * [\log\{K_t\}]) + 0.0724\,DUM1 + 0.0262\,TREND$$
$$+0.00026\,RAINFALL + 0.4276\,AR(1)$$

$$Y_t = \exp(Log(Y_t))$$

The time period considered for the purpose of simulation is 1951 – 2000, i.e., the sample period. In other words, all the counterfactual simulation models are solved for the above time period. The model is solved for the above mentioned period. The actual values of output and the forecasted values of output obtained by solving the model are given below:

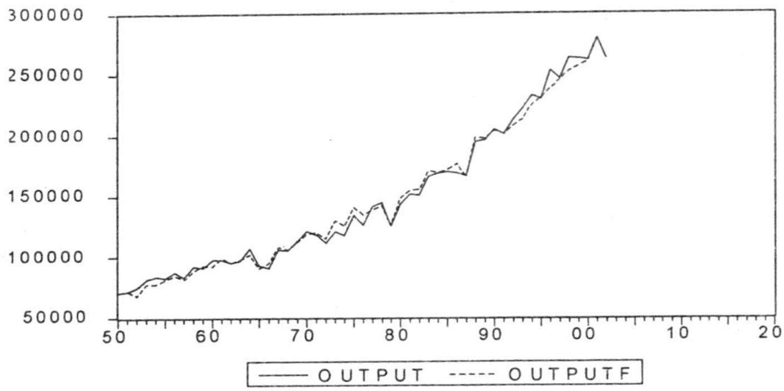

A look at the graph above indicates that the model is able to predict the actual values very well. Using the forecasted values as baseline, a series of counterfactual simulation exercises have been done below, under different set of assumptions.

Counterfactual Simulation Exercise 1

In this exercise, we bring out the importance of rainfall and other natural factors in influencing the overall agricultural output. We go on to show that Indian agriculture is still largely 'rainfall dependant'. To carry out this exercise, we remove the variable RAINFALL and the drought dummy from the basic model and then see the resulting output. This output is compared with the baseline values to identify the difference.

The model that is solved under this exercise is–

$$LOG(Y_t) = 10.96236 + 7.8571\, D(\log\{L_t\}) + 1.4527\, D(D(\log\{K_t\}))$$

$$-0.6140\, D([\log\{L_t\}] * [\log\{K_t\}]) + 0.0262\, TREND$$

$$+0.4276 AR(1)$$

$$Y_t = \exp(Log(Y_t))$$

The output values that are obtained by solving the above model are compared with the baseline values in the graph given below.

The series with the dotted lines indicates the level of agricultural output without the variable RAINFALL. As seen,

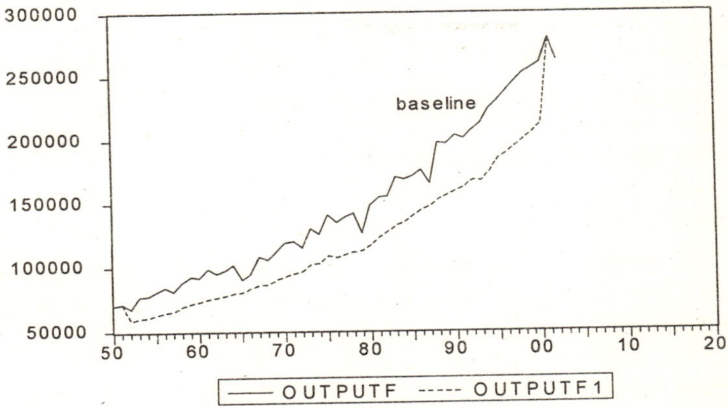

it is lower than the baseline values of output that include RAINFALL. It only indicates the importance of rainfall in influencing the overall agricultural output in India Another feature that is observed is that, the gap between the baseline values and the simulated values is widening continuously over the time period concerned. It is only indicative of the fact that Indian agriculture may have become more rainfall dependant over time. And this is in spite of large scale improvements in the irrigation facilities over time. Finally, by removing the drought dummy, we find that the new series is no longer able to capture the major drought years namely, 1966, 1979, and

1987. But these drought years are well captured by the baseline series as indicated by the 3 relatively large troughs in the series.

Counterfactual Simulation Exercise 2

In this exercise we bring out the importance of technological change in influencing the agricultural output. Here, we consider mainly the neutral technological change and not embodied technological change. We carry out this exercise by removing the variable TREND from the basic model. The significant trend coefficient in the model is indicative of the existence of neutral technological change in the Indian agriculture. It also measures the impact of other factors not explicitly considered in the production function. The model that is solved in this simulation exercise is given as follows-

$$LOG(Y_t) = 10.96236 + 7.8571\, D(\log\{L_t\}) + 1.4527\, D(D(\log\{K_t\}))$$

$$- 0.6140\, D([\log\{L_t\}] * [\log\{K_t\}]) + 0.0724\, DUM1$$

$$+ 0.00026\, RAINFALL + 0.4276\, AR(1)$$

$$Y_t = \exp(Log(Y_t))$$

The output values that are obtained by solving the above model are compared with the baseline values in the graph given below.

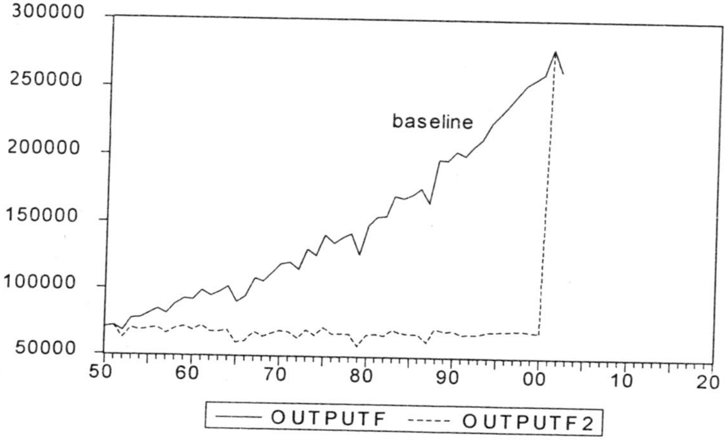

In the above graph, the series with the dotted lines is the simulated series. As can be seen, without the impact of neutral technological change and other factors not explicitly considered in the production function, the output level is almost constant. In fact it has marginally reduced. The conclusion that is clearly established is that, a very significant part of the increase in agricultural output during the period 1951-2000 is attributed to 'technological change', without which, this much increase in output would not have been possible.

Counterfactual Simulation Exercise 3

We have earlier calculated the marginal products for capital and found them to be positive all through and increasing from the decade 1971-80 onwards. Given this fact, an increase in the amount of capital at any given time period should have a positive impact on output. To verify this fact, we conduct 2 counterfactual simulation exercises. We consider 2 arbitrary time periods 1981-82 and 1991-92. At each of these time periods, we raise the amount of capital by a specific amount. Again, this increase in capital can be of two types–

(a) One Period Shock

(b) Multi-Period Shock

In the first case, the level of capital is changed (increased or decreased) only for one particular chosen time period. The level of capital for the other time periods remains the same as before. In the second case, the level of capital is changed by a fixed amount for all the time periods following a particular chosen time period.

In this simulation exercise, we apply a one period shock on capital for the time periods 1981-82 and 1991-92 separately and see the corresponding impact on output.

First, we consider the time period 1981-82 and give one period shocks for capital during this time period. We raise the level of capital by different amounts and compare their impacts on the output. Given that the variable Capital is measured in 'thousands of rupees', we consider the following 4 different one-period shocks namely, raising the amounts of capital by 1000, 10000, 20000, and 40000 respectively. The impacts of these shocks on output are shown in the graphs given below. The first graph shows the effect on output of different magnitudes of one period shocks given during the year 1981-82. The second graph shows the effects on output, for the same shocks given during the year 1991-92.

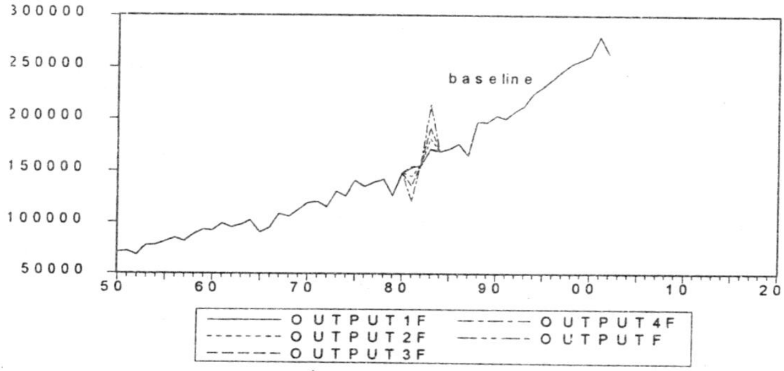

As seen from the above two graphs, before the time period of the shock, all the output levels coincide with the baseline. During the period of shock there is a fall in the level of output relative to the baseline. The magnitude of the fall in output is directly proportional to the magnitude of shock given to the variable capital. In other words, greater the magnitude of shock given to capital, greater is the fall in output during the period of shock and vice versa. In the two periods that immediately follow the period of shock, the output levels show an increase relative to the baseline. The levels of output during these two periods are directly proportional to the magnitude of shock given. After this, the levels of outputs, irrespective of the magnitude of shock given, once again coincide with the baseline for the remaining time periods. To sum up, the effect of a one period shock is felt for a total of 3 consecutive time periods. During the period of shock, the output level falls, and in the following two consecutive time periods, it rises. After this, the output levels coincide with the baseline.

An important conclusion that can be drawn from the above two graphs is that, the behaviour of output in response to a given magnitude of shock, is more or less same, irrespective of when the shock is administered.

Counterfactual Simulation Exercise 4

In this simulation exercise, we administer a multi–period shock on the variable capital and see its impact on the levels of output. Here also we consider two time periods, 1981-82 and 1991-92 and administer separately multi–period shocks starting from the above respective time periods.

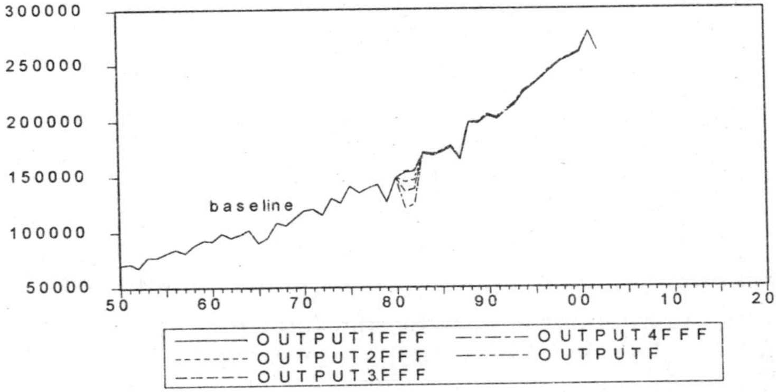

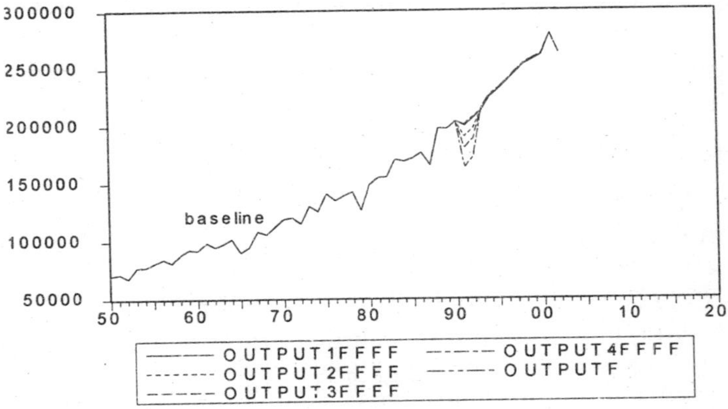

First, we consider the time period 1981-82 and give multi-period shocks for capital starting from this time period. We raise the level of capital uniformly from this time period by different amounts and compare their impacts on the output. We consider the following 4 different multi-period shocks namely, raising the amounts of capital by 1000, 10000, 20000, and 40000 respectively. The impacts of these shocks on output are shown in the graphs given below. The first graph shows the effect on output of different magnitudes of multi-period shocks given from the year 1981-82. The second graph shows the effects on output, for the same shocks given from the year 1991-92.

As seen from the above two graphs, before the time period of the shock, all the output levels coincide with the baseline. During the first period of shock, there is a fall in the level of output relative to the baseline. The magnitude of the fall in output is directly proportional to the magnitude of shock given to the variable, capital. In other words, greater the magnitude of shock given to capital, greater is the fall in output during the first period of shock and vice versa. From the second time period of the shock onwards, the output levels rise, and this rise in the levels of output continues thereafter for all the subsequent time periods. Even though the output levels rise during the second period of the shock, they are lower than the baseline values. In all the subsequent time periods, the output levels are higher than the baseline values and the levels of output are directly proportional to the magnitude of shock.

As with the case of one period shocks, the behaviour of output in response to a given magnitude of shock, is more or less same, irrespective of from when the shock is administered.

CONCLUSIONS AND POLICY IMPLICATIONS

The objective of this study is to identify and examine the technological change in Indian agriculture and analyze its impact on the factor-factor relations. The sample period for the study is 1950-51 to 1999-00. Initially a production function is fit with the main inputs being land and capital. Other variables such as, labor, irrigation and fertilizers were also considered, but due to inherent problems in the nature of these variables, they had to be excluded from the production function. Rainfall was included in the production function. A trend variable was also included in the production function. It is expected to represent neutral technological change along with the effect of factors not included in the analysis. They would include also such variables as infrastructures, HYVs, fertilizers etc. Its coefficient is highly statistically significant and its value indicates a high rate of neutral technological change over the five decades under study.

Next, the marginal products of the two inputs were estimated for the five decades spanning the period 1951-2000. The marginal products of land were found to be increasing over the period of five decades. The marginal product of capital decreased initially, but from the decade 1971-80 onwards, it was increasing. The factor-factor relationship among land and capital was also estimated using the Allen's partial elasticities of substitution for the five decades spanning the period 1950-1999. The results suggest a substitution relationship between land and capital all through the period under consideration. These results could be explained only in terms of the likelihood of embodied technological change in capital over the three decades, spanning the period 1970-1999.

Even though there is some indication of technological change as embodied in capital, the major change in production was due to neutral technological change. The period 1970-99 is the period marked by the impact of Green Revolution, potent with a high degree of technological change.

As shown above, two major features of our investigation prominently stand out from the policy point of view. In the first place, given the land input, the marginal product of capital in the recent decade has shown a general improvement over the preceding two decades. This has, however, been accompanied by a substantial increase in the volume of capital input. This apparently unexpected result is important because, the only way it can be reconciled with the celebrated principle of diminishing marginal product is that, capital incorporates a substantial measure of embodied technological change. For policy purposes, it is therefore important to view increased investment in agriculture as carrying with it elements of embodied technology. This brings into focus the role of tractors, harvesters, power tillers and oil engines, electric motors and machinery which make up capital on farm. Our study also brings out the fact that a one time increase in the amount of capital will have only a limited and short run impact on the agricultural output. On the other hand a multi-period increase in the amount of capital will have a greater and a long–run impact on the agricultural output.

We also observe increased marginal product of land, which may result in increased value of land. If, therefore, land is used as equity, credit level may increase and that may facilitate additional capital formation. For this to happen, appropriate institutions are a prerequisite.

Our results also support the view that neutral technological change has made a very significant and sizeable contribution to the agricultural output. Neutral technology may contain the impact of fertilizer use, HYV and pesticides. We infer that it could be also the result of the spread effect mentioned earlier. The new technology might be emerging to cover the hitherto uncovered crops, and regions uncovered or poorly covered by the new technology. If this trend in technology spread is consciously encouraged as a policy measure, the economy could be back on the dynamic growth path over the coming decades.

REFERENCES

Desai, Bhupat M. (1994), "Contributions of Institutional Credit, Self-Finance and Technological Change to Agricultural Growth in India", *Indian Journal of Agricultural Economics,* Vol. 49, No. 3, July-September.

Farmer B. H. (1986), "Perspectives on the 'Green Revolution' in South Asia", *Modern Asian Studies,* Vol. 20, No. 1.

Feder Gershon, Just E. Richard, David Zilberman (1985), "Adoption of Agricultural Innovations in Developing Countries: A Survey", *Economic Development and Cultural Change,* Vol. 33, No. 2, Jan.

Goldar, B. N. and H. Mukhopadhyay (1991), "Substitution Among Labour, Capital And Energy In Indian Industries: An Application of the Cost-Price Approach", *Journal of Quantitative Economics,* Vol. 7, No. 1 (Jan).

Jha. R., M.N. Murty and S Paul (1991), "Technological Change, Factor Substitution and Economies of Scale in Selected Manufacturing Industries in India", *Journal of Quantitative Economics,* Vol. 7, No. 1 (Jan).

Kawagoe Toshihiko, Keijiro Otsuka, and Yujiro Hayami (1986), "Induced Bias of Technical Change in Agriculture: The United

States and Japan", *Journal of Political Economy,* Vol. 94, No. 3 (June).

Krishnamurty K., V. Pandit and G. Mahanty (2004), 'Policies for Stability and Growth (Experiments with Large and Comprehensive Structural Model for India)', in, V. Pandit and K. Krishnamurty (eds), *'Economic Policy Modelling for India'*, Oxford University Press, New Delhi.

Kuroda Yoshimi (1988), "The Output Bias of Technological Change in Postwar Japanese Agriculture", *American Journal of Agricultural Economics,* Vol. 70, No. 3, (Aug).

Mehra Shakuntala (1981), *"Instability in Indian Agriculture in the Context of the New Technology"*, Research Report 25, IFPRI, July.

Mruthyunjaya, Suresh Pal, P. K. Joshi and Raka Saxena (2006), 'Institutional Reforms for Agricultural Development under New Economic and Technological Environment', in, N.A. Mujumdar and Uma Kapila (eds), *'Indian Agriculture in the New Millennium, Changing Perceptions and Development Policy'*, Academic Foundation, New Delhi.

Nerlove, Marc (1996), 'Reflections on Agricultural Development, Population Growth, and Environment', in, M.G. Quibria and J. Malcolm Dowling (eds), *'Current Issues in Economic Development-An Asian Perspective'*, Oxford University Press, Oxford, New York, Hong Kong.

Nkamleu Guy Blaise (2004), "Productivity Growth, Technical Progress and Efficiency Change in African Agriculture", *African Development Review,* Vol. 16, No. 1 (Apr).

Rao, C. H. H. (1975), *"Technological Change and Distribution of Gains in Indian Agriculture"*, Macmillan and Co., Delhi.

Reserve Bank of India, *"Handbook of Statistics on the Indian Economy 2004-05"*.

Schultz T. W. (1964), *'Transforming Traditional Agriculture'* Yale University Press, New Haven, Connecticut, p. xiv, 212.

Sivasubramonian S. (2004), *"The Sources of Economic Growth in India 1950-1 to 1999-2000"*, Oxford University Press, New Delhi.

Srivastava Uma K., Earl. O. Heady (1973), "Technological Change and Relative Factor Shares in Indian Agriculture, *American Journal of Agricultural Economics,* Vol. 55, No. 3, August.

Stevenson R. (1980), "Measuring Technological Bias", *American Economic Review,* Vol . 70.

12

TECHNICAL AND ALLOCATIVE EFFICIENCY IN AQUATIC FARM PRODUCTION: EMPIRICAL EVIDENCE FROM SELECTED DISTRICTS OF ORISSA

Surath Nayak

INTRODUCTION

Efficiency of a farm refers to its performance in the utilization of resources at its disposal and is a relative concept. It is measured either with respect to the normatively desired performance of a farm. Hence the efficiency measures are basically the methods of comparing the observed performance of a farm with some specified performance. The economic efficiency combines allocative and technical efficiencies. A farm is said to be more efficient technically than other if it is consistently produces larger quantities of output from the same quantities of measurable inputs. The differences in technical efficiency among group of farms can therefore be captured by quantifying their differences in "Technology". On the other hand, a farm is said to be price efficient if it maximizes its profit. As profit maximization implies equalization of the value of marginal product of each variable input to its price, analysis of price or allocative efficiency also includes prices as variables that are not ordinarily included in production function analysis. Difference in productive efficiency among group of firms may be caused by difference in technical and allocative efficiency suggested by Hoope, Schultz and Yotopoulus (1967).

Lau and Yotopoulus (1971) in their paper states that, the relative productive efficiency of two farms can be compared with varying degrees of technical and allocative efficiency. For this purpose the actual profit functions of the two firms are compared at given output and input prices and quantities of measurable fixed inputs. The firm which has higher profit, that is derived from total revenue and total cost of the variable factors of the production function within a certain specified range of output and input prices is considered to be more economic efficient within the range of prices. By applying the concept in comparison of small and large type of farms, it is observed that small farms are more economic efficient. For many purposes, it is important to assess the relative importance of the two factors responsible for observable differences in relative productive efficiency.

Henderson and Quandt (1971) in their study stated that, a production function presupposes technical efficiency and states the maximum output obtainable from every possible input combination. Allocative efficiency is said to exist if "the entrepreneur may desire to maximize his output level for a given cost or may desire to minimize the cost of producing a given output level". The first order conditions for both the problems require that the rate of technical substitution between the inputs be equated to their price ratio. The entrepreneur may allow both output level and cost to maximize his profit. First order conditions require that the value of the marginal physical productivity of each input be equated to its prices. Consider a farm that uses 'n' inputs $Y = F(X_1, X_2, X_3 \ldots\ldots\ldots\ldots X_m)$ available at fixed prices $W = (W_1, W_2, W_3 \ldots\ldots\ldots\ldots Wn)$ to produce single output Y. The farms technology is characterized by the production function f(X), which shows the maximum output obtainable from various input vectors. If a farm produces Y^0 level out output using X^0 inputs, it is said to be technically efficient if $Y^0 = f(X^0)$ and technically inefficient if $Y^0 \leq f(X^0)$ thus the existence of technical inefficiency. So that productivity of at least one input is lower than what it would be without technical inefficiency. Whether it reduces productivity of all the inputs or not depends on the functional from of the technology. For example, with a Cobb-Douglas technology technical inefficiency reduces productivity of all the inputs where as with a linear production function; it might affect only some particular input. Again even in the Cobb-Douglas model, there is no reason to believe that productivity of all input will reduce equi-proportionately. The technical inefficiency measure was introduced by Aigner et all (1977) and Meensen and Vanden Broeck (1977) implies an equi-proportionate reduction in marginal productivity of all the inputs which is restrictive. Such an efficiency formulation is called aggregative since there is no way to identify specific to each input. The necessity of input specific technical inefficiency arises because of the fact that knowing the magnitude of technical inefficiency is not enough. It is more important to know which input is causing the inefficiency and to what extent. Such type of inefficiencies can be measured with the use of allocative efficiency, either by using the method of first order of cost minimization or by profit maximization method. The advantage of a disaggregate measure is that it not only identifies the inputs causing inefficiency but also the contribution of such input specific inefficiency to the aggregate level of technical inefficiency for the farm.

In order to measure technical efficiency, a specified standard is required against which the observed performance of the firm can be measured. Production frontier serves as one such standard in the case of technical efficiency, which is the outer boundary in output space for a given set of input points. So the technical efficiency is defined as the amount by which the actual output fall short of the maximum possible output (Ramaswamy, 1994). Attempts to estimate technical efficiency using frontier began with "non-stochastic", nonparametric and non-statistical frontier model approach pioneered by Farell (1957). Following the initiative by Farell the estimation of 'Non-stochastic parametric non-statistical frontier model began with the study of Aigner and Chau (1968). Non-stochastic parametric statistical production frontier model' was first explicitly specified by Afrait(1972). Richmond (1974) has apparently noted the correction in Ordinary Least Square (OLS) estimates, subsequently the non-stochastic parametric approach is discussed by Schmidt (1976), Lee and Tyler (1979), Greene (1980), Bhavani (1991), Sharma and Sinha(1995). Theoretically the concept of production frontier is nothing but production function. By definition, production function represents the maximum possible output for any given set of inputs, so it sets a limit or frontier on the observed values of output that is expected to lie above the production function. It is therefore refereed a production frontier or frontier function. The word frontier represents the concept of optimality. As the production frontier represents maximum obtainable output for any given set of input and represents the degree of technical inefficiency. The usual estimation of production function by OLS assumes that error term 'U' can be positive and negative with zero expectation' with the definition the production function,

$$Y = f(X)\, e^{-u}$$

Where e is the multiplicative error term and can be taken as the ratio of observed output (Y) to potential output f(X), and $e^{-u} = Y/f(X)$. So the usual multiplicative error term can be taken as measure of technical efficiency. As the production function is expected to represent maximum possible output. The observed output (Y) would always be less than or equal to potential output f(X). In log form, $\ln Y = \ln f(X) - u$ where u takes the value between zero and infinity. The appropriate way of estimating production function is to treat u as a random variable with $(0, \infty)$ range and hence drawn from a one-sided statistical distribution. In this way production function (ln Y) represents frontier as given by the definition. Allocative efficiency or price efficiency refers to the selection of optimum combination of inputs which recognizes input prices and marginal products. The stochastic production frontier model is $\ln y = \ln \alpha_0 + \Sigma\, \alpha_i \ln x_i + v - u$ while the non-stochastic production frontier model is Ln y = ln $\alpha_0 + \Sigma\ \alpha_i \ln x_i - u$. If the model assume to operate under behavioral assumption of cost minimization and constrained profit maximization, allocative inefficiency can be modelled by permitting the cost minimizing conditions, which define the least cost expansion path in implicit from to fail to hold exactly. It may be allowed for the possibility that the firm my also be allocative inefficient by permitting it to operate off its least cost expansion path. The error in choosing cost minimizing factor proportions then corresponds to disturbance from the exact satisfaction of the first order condition for cost minimization.

$$\text{So, } \ln (x_1 / x_i) = \ln (\alpha_1 w_1 / \alpha_i w_i) + e_i$$

Where i = 1, 2. 3. .... n x_1 is the capital input and αi_s are the inputs i.e., labour for any inputs and so on. Here e_i represents the amount by which the ith first order condition for the cost minimization fails to hold. The parameters $\alpha 1$ and α_i (i= 1, 2., 3....n) are the estimates derived from the production function. All α_i are same in the classical average function and in both the frontiers estimated by correcting OLS methods. Both the frontiers estimated by the corrected ordinary least square and moment method. In India past studies mainly focused on testing whether resources are allocated optimally or not. In context, this allocative hypothesis (Schultz, 1964), is tested by using production function method and more recently applying stochastic and deterministic frontier production function and profit function method.

An attempt is made to study both technical and allocative effect of education of the farm people in different size of holding by using farm level data for two consecutive years. Here the production function method is used to estimate the economic contribution of schooling of the "effective farm household" on efficiency in farm production. The production function is estimated holding-wise and farm business as a whole. The productive value of education has its root in two distinct phenomena. Increased education simply may permit a worker to accomplish more with resource at hand. This worker effect is the marginal product of education as it is normally defined as the increased output per unit change in education holding other factors constant. On the other hand, increased education may enhance a worker's ability to acquire and decode information about cost and productive characteristic of other inputs. As such change in education results in a change in other inputs including perhaps the use of some new factors that otherwise would not be used. The return to education is therefore considered as consisting of two effects i.e., workers effect and allocative effect. The study exploring the economic effects of education (Grilliches, 1964) used production function analysis to highlight the contribution of education in farm productivity. The first clear-cut distinction between worker effect and allocative effect is provided by Choudhury (1968). He is trying to assess the impact of education on Indian agriculture and estimate an aggregate production function at the state level. His conclusion reinforces the Nelson and Phelps (1966) study that education enhances innovative ability and the use of modern input positively increases the farm production. The study by Welch (1970) shows that in Cobb-Douglas type of production function for single crop and multi crop, the output is measured in terms of gross sales value and a value added production function can be used to distinguish the effect of technical education on farm production. In each case the production is assumed to be technically efficient in a sense that for a given inputs, the physical output is maximized.

The allocative effect on education can be measured by estimating a production function for values of all crops produced by a farm. The two component of allocative effect can be measured by fitting two production functions such as value added production function in which the independent variable does not include the value of variable inputs and a gross value production function which includes both the values of fixed and variable inputs. However, due to some difficulties in computing the shares of farm supplied inputs and purchased inputs the value added function is not taken into consideration. Hence, the input allocation part of allocative efficiency is considered. The gross value function is given in this form. $Y = g(X, E)$ where Y is the gross value of output, X is the vector of fixed and variable inputs, and E is the vector of education. The maximization of y requires both efficient allocation and optimum utilization of given resources. If it is assumed that the allocation of X among its competing uses is a function of the vector E, the marginal value product of education captures both allocative and technical efficiency.

DATA SOURCE

The data used in this study are primary and were collected during 2004 -06, approximately twenty four months. During this period the complete cultivation takes place. The area covers two blocks/taluks of Puri district, which consists of nine revenue villages. This block is selected with the objective that massive changes takes place in changing the occupation from shifting paddy cultivation to prawn cultivation. The households under consideration were selected on the basis of random sampling with the help of readymade questionnaire made in both English and Oriya language. The data for some aspects were also collected from the District Agricultural offices and Directorate of Economics and Statistics, Bhubaneswar. Two hundred farms have been surveyed to get detail information regarding inputs used, input prices and output quantity and prices, family background and educational status of the people working in farms. The data relating to the cultivation and other logistic aspects were also collected from taluk offices, which are relevant for the study. The questionnaire divided into eight parts covers the information relating to socio-economic aspects of the farm households, production and productivity, area under the cultivation, inputs (price and quantity) and determining the value of output excluding depreciation charges and value of the machine used.

METHODOLOGY

Empirical study shows that there are three approaches in estimating technical and allocative effects of education or skill i.e., (i) Production function approach, (ii) Cost function approach and (iii) profit function approach. As the objective of the study is to understand the effects on output not profit, the Cobb-Douglas type of production function is chosen. The production function for each crop with output measured in Physical quantity as well as whole farm production function in terms of gross value are used. The estimate of both production functions are obtained through Ordinary Least Square (OLS) methods.

Economic efficiency extends this concept to take into account the relative cost of alternative inputs as well. A production process is called economically efficient if the inputs are combined to produce given level of output at minimum cost. Many scholars have applied this concept in relative terms, on the basis of argument that efficiency improvements requires reducing the total cost of inputs used to produce a given level of output. In other words, if the total cost of inputs used by the process X to produce a given level of output is less than the cost of inputs required by another process Y to produce the same output, then the process X is more economically efficient than the process Y. The optimal input-output combination is given by the tangency point between the iso-quant and iso-cost (price) line. Thus economic efficiency describes a specific subject of technically efficient choices. In fact for a given set of inputs, prices and output levels there can be only economically efficient combination of inputs, although more technically efficient alternative are possible. The statistical approach makes use of the production function in order to estimate or otherwise characterize the input-output relationship in the prawn farms. The shape of the production function is usually specified a priori, on the basis of economic theory and practical consideration of simplicity. Similar procedure is also followed for cost function estimation, which quite often is used instead of production function to deal with multiple outputs and inputs. Thus the production function is used to deal explicitly with multiple inputs and single output. Let us examine a production function in a specified form where output is a function of many inputs. For example,

$$Y = F(X_1, X_2, X_3 \ldots\ldots\ldots\ldots X_m)$$

Where, X_1, X_2, and X3 ....Xm represent quantity/ value of each inputs and Y is the corresponding output. The functional form "F" is assumed to be known as each output value Y is maximum in order to represent the above produ0ction function. Given the situation we may distinguish the observed input amount for the same farm by indexing them as X_1j, X_2j, X_3j X_3j .... Xmj, So that Xij represents the amount of 'i'th input used by the farm j using these inputs, the "J" th farm producing the output Yj, which satisfies the condition that

$$0 \leq Y_j \leq Y^*_j$$

where, Yj* = f (X_1j, X_2j, X_3j ...mj)

In other words Yj* represent the maximal output, which could have been achieved if the farms had used the input efficiency. For simplicity the equation (2) can be written in the ratio form i.e., ≤Yj/Yj*≤1 with an efficiency ratio of unit being achieved only if the production process is fully efficient. Failure to achieve ratio unit would then represent what might be termed as managerial or technical inefficiency irrespective of prices paid for these inputs. Of course management which achieves Yj* but paid excessive price for the inputs would become economically efficient. This represents price inefficiency which is closely related with the concept of price inefficiency close to allocative inefficiency. Even if the prices paid are minimum, when the farm does not respond to the opportunities the market allow for the substituting between factors in order to achieve a level of maximum output level Yj*. In other words mix of inputs is used to achieve this output.

Multiple regression analysis approach is used to estimate the input-output relationship. Specification of input units in a specific functional form requires understanding a theoretical background, which is necessary to understand the flexibility of the model. When the relationship among the variables (inputs) seems elastic in nature, then the best method to consider Cobb-Douglas type of production function which clearly explain the degree of relationship between dependent and independent variable with a given technology can be specified in the following form,

$$Y = A X_1^{\alpha} X_2^{\beta}$$

A, α, β ≥ 0 are parameters, where A is constant and α, β represent output elasticity with respect to factor inputs X_1 and X_2 respectively. The production process exhibits, increasing, decreasing and constant returns to scale, when $\alpha+\beta>1$, $\alpha+\beta<1$ and $\alpha+\beta=1$ respectively. However the above equation can be estimated under logarithm form, which can be expressed as,

$$\text{Ln}\, Y = \ln A + \alpha \ln X_1 + \beta \ln X_2$$

The above equation is usual form of fitting the Cobb-Douglas production function by the method of Ordinary Least Square (OLS) to determine the value of A, a, b. Statistical test is applied to determine the significance of the model. The above function can be extended in the case of more than two inputs by expressing in the following form,

$$\text{Ln}\, Y = \ln A + \Sigma\, \text{Ln}\, \alpha_i\, \beta \ln X_i$$

Where, i = 1,2,3 ....n

The same model can be used to determine the factors responsible for prawn culture. Which is discussed under the factors which determine the growth of prawn culture.

SPECIFICATION OF THE MODEL

In prawn farming, the production depends on the factors like size of the tank, stock of seeds, seed survival rate, amount of feed, fertilizers, fuel, labour used in shrimp culture. Therefore, it is necessary to examine the allocative efficiency of factor inputs and its share in production of aquatic products. In order to examine the effect of education/skill and other inputs on output, a Cobb-Douglas type of production function can be considered. The estimation of shares/elasticity can be done through simple Ordinary Least Squares (OLS) method. The functional form of the model is given below.

Output (Y) = F (Fertilizers, Seed stock, size of the tank, rate of survival, Quantity of feed used, Manual labour , Power pumps and skilled labour other variables).

The linear regression equation can be expressed in the following form.

The required model in functional form;

$$Y = a\, X_1^{\beta 1}\, X_2^{\beta 2}\, X_3^{\beta 3}\, X_4^{\beta 4}\, X_5^{\beta 5} X_6^{\beta 6} X_7^{\beta 7}\, X_8^{\beta 8}\, U_t.$$

The linear regression equation:

$\ln Y = \alpha + \beta_1 \ln X + \beta_2 \ln X_2 + \beta_3 \ln X_3 + \beta_4 \ln X_4 + \beta_5 \ln X_5 + \beta_6 \ln X_6 + \beta_7 \ln X_7 + \beta_8 \ln X_7 + D$

Let us assume that,

$\beta_1 > 0, \beta_2 > 0, \beta_3 > 0, \beta_{:4} > 0, \beta_5 > 0, \beta_6 > 0, \beta_7 > 0$ and $\beta_8 > 0$

The assumption regarding coefficients can be verified with results.

Where,

Y = Output in kg. Per- crop,

X_1 = Fertilizer used in kg per crop/per Hect.

X_2 = seed stock in numbers per crop/per Hect.

X_3 = Size of the tank/pond per thousand square ft.

X_4 = Rate of survival of the prawn seed per thousand.

X_5 = Quantity oı feed used in kg. Per crop/per pond

X_6 = Amount of labour used in hours/per pond/per Hect.

X_7 = Power Pump used per hour/per pond

X_8 = Level of education/skill.

D = 1, if the crop is success (Dummy)

D_2 = 0, if otherwise, (Dummy)

And α = Constant, $\beta_1, \beta_2, \beta_3$ .... β_8 are regression coefficients/parameters.

The functional form is estimated by the method of Ordinary Least Squares (OLS) using the crop-wise data of sample farms/ponds. The Cobb-Douglas model has been used to estimate separately the share of factor inputs and its elasticity for different size of holding. For the study convenience the holding size of the ponds has been classified into three categories-the holding size-I for small and marginal holdings, holding size-II for medium type of holding and holding size-III for large size.

RESULTS AND DISCUSSION

Let us examine the model in the case of holding size-I. In the case of the small and marginal size holding, the estimated model is given below. The estimated regression equation is,

$$Y = 0.8179 + 0.4975\ X + 0.6433\ X_2 + 0.4532\ X_3 + 0.6211\ X_4 - 0.4980\ X_5$$
$$(6.3471)^{*}\ (0.3257)^{**}\ (0.5481)\ (2.8566)^{*}\ (1.7329)^{**}$$
$$-\ 0.3255\ X_6 + 0.4177\ X_7 + 0.5893\ X_8 + 0.04431\ D_1 - 0.0077\ D_2$$
$$(0.5572)^{*}\ (2.9855)\ (3.5433)^{**}\ (0.0769)\ (0.1538)$$

$R^2 = 0.9162$ $R^{-2} = 0.8975$ D.W = 2.5433 SEE = 0.6732

*: Values given in the bracket are t- values.

In the above result it is clear that the variation (in the dependent variable can be explained to the extent of 91% by the variables used in the model. The use of fertilizer, size of tank, rate of survival, quantity of feed used and amount of labour used per hours, level of education, and seed stocks have significant effect on the potential realization of aquatic yield. The elasticity coefficient for fertilizer use (0.4983), rate of survival (0.6221), quantity of feed used (0.4980) and amount of labour used per hour (0.4117) is positive and significant at 1% level. However the elasticity coefficient in case of Seed stocks (0.6433) and quantity of feed used (0.4980) are positive and significant at 5% level. Finally the elasticity of the level of education of the labourer is 0.5893 showing positive relation and significant at 10% level. It is because in case of small and marginal holding the level of education is not is so important than other factors. With respect to dummy variable it is also found that the degree of elasticity is insignificant. The value of the standard error is also minimum in the above estimation. The R^{-2} value shows a goodness-of-fit in explaining the relationship between dependent and explanatory variable. The value of the Durbin-Watson statistics shows that there is no auto-correlation among the variables (Table 1).

The regression analysis of holding size-II i.e., for medium holdings revealed that the factors namely fertilizers, seed stocks, rate of survival, amount of labour used and level of education is important in determining the potential yield of aquaculture. In this type of holding the dependent variable is

Table 1
Estimation of Regression Coefficients for Holding Size–I

Independent Variable.	*Elasticity*	*Stand. Error*	*t– value*	*Mean (S.D)*
Intercept	0.8179	1.0437	1.9843	217.32 (39.97)
Fertilizer used (kg/per pond)	0.4975	0.9982	6.3471*	14.117 (1.256)
Seed stocks (No/per crop)	0.6433	1.0953	0.3257**	34.731 (17.39)
Size of the tank	0.4532	1.0639	0.5481	592.37 (30.11)
Rate of survival (per 1000)	0.6221	2.4337	2.8566*	191.32 (36.12)
Quantity of feed used (kg/per pond)	0.4980	1.7344	1.7329**	186.49 (87.09)
Electricity consumption	-0.3255	0.5499	0.5572	9.232 (5.167)
Amount of labour used in hours	0.4177	0.6655	2.9855*	8.135 (2.391)
Level of education/ skill of labour	0.5893	1.8774	3.5433**	12.572 (2.519)
Success rate (D_1: Dummy)	0.0443	0.3277	0.0769	0.912 (0.568)
Failure or otherwise (D_2: Dummy)	-0.0063	0.3241	- 0. 1538	0.1510 (0.379)

$R^2 = 0.9162$ $\bar{R}^2 = 0.8975$ S.E.E = 1.4329 DW = 2.5433

Source: See Annexure-III. (Field work survey report)
Note: *: Significant at 1 % level
**: Significant at 5% level

explained to the extent of 87% by the independent variables. Here the size of the farm is very conducive to protect and preserve the seeds for long time. The elasticity coefficients for fertilizer used per kg (0.5975), seed stocks (0.6433), rate of survival (0.7224), and amount of labour used (0.5177) are positive and significant at 1% level. However, the elasticity coefficient for the quantity of feed used in kg/per pond (0.5983) and level of education (0.4923) are positive and significant at 5% level. In this type of holding the success rate of dummy variable has positive implication for the total output. The R^{-2} value is 83%. The standard error of the equation is also minimum (Table 2). Hence the above-mentioned variables have significant influence over dependent variable. The estimated regression equation for the size of holding II (medium size holding) is given below.

$$Y = 0.9537 + 0.5975\,X_1 + 0.6433\,X_2 + 0.3211\,X_3 + 0.7224\,X_4 + 0.5983\,X_5$$
$$(2.3471)^{*}\ (3.6257)^{*}\ (2.7321)^{*}\ (2.4566)^{*}\ (1.5732)^{**}$$

$$+ 0.3255\,X_6 + 0.3177\,X_7 + 0.4293\,X_8 + 0.0542\,D_1 - 0.0074\,D_2$$
$$(0.8572)\ (2.9855)\ (3.5432)\ (0.0769)\ (0.0157)$$

$R^2 = 0.8797$, $\bar{R}^2 = 0.8397$ S E E = 0.8544 D. W = 4.6937

* The values given in the bracket is t-values.

Table 2
Estimation of Regression Coefficients for Holding Size–II

Independent Variable.	*Elasticity*	*Stand. Error*	*t – value*	*Mean (S.D)*
Intercept	0.9537	0.5422	1.0955	397.4 (51.72)
Fertilizer used (kg/per pond)	0.5975	0.5437	7.3471*	2.359 (0.951)
Seed stocks (No/per crop)	0.6433	1.7655	3.6257**	91.275 (35.33)
Size of Tank/Pond	0.3211	0.6539	2.7321	126.49 (11.98)
Rate of survival (per 1000)	0.7224	0.7785	2.4566*	13.371 (49.55)
Quantity of feed used (kg/per pond)	0.5983	1.6548	1.5732**	27.65 (18.54)
Power consumption	0.3255	0.6653	0.8572	91.37 (36.57)
Amount of labour used in hours	0.3177	0.7942	2.9855*	18.511 (2.366)
Level of education/ skill of labour	0.4293	0.4317	3.5432**	11.573 (2.036)
Success rate (D_1: Dummy)	0.0542	0.4359	0.0769	0.5771 (0.31)
Failure or otherwise (D_2: Dummy)	-0.0074	0.7863	- 0.0157	0.9128 (0.13)

$R^2 = 0.8729$, $\bar{R}^2 = 0.8397$ S E.E = 0. 8544 DW =4.6937

Source: See Annexure-III

Note: *: Significant at 1 % level

**: Significant at 5% level

Similarly, in the case of large holding i.e., holding size III, it has been noticed that the dependent variable is explained to the extent of 92% by the independent variables. The variables like seed stocks, rate of survival, quantity of feed used and level of education of the workers have significant effect on the potential realization of the yield. The elasticity coefficient of fertilizer used per kg (0.7985), rate of survival (0.6227), amount of labour used (0.4117) and level of education (0.5893) are positive and significant at 1% level. However the factors like seed stocks (0.6433) and quantity of feed used kg/per pond (0.4980) are positive and significant at 5% level. The adjusted R^2 value is slightly smaller than R^2 value. The standard error is also minimum. The Durbin Watson statistics shows that there is no auto-correlation among the variables (Table 3). The estimated regression equation for the large size holding is given below,

$$Y = 0.9172 + 0.7985\, X_1 + 0.6433\, X_2 + 0.8732\, X_3 + 0.6227\, X_4 + 0.4983\, X_5$$

$$(4.3471)^{*}\ (0.3255)^{**}\ (0.4327)\ (2.4562)^{*}\ (1.7325)^{**}$$

$$+ 0.3255\, X_6 + 0.4117\, X_7 + 0.5893\, X_8 + 0.0443\, D_1 - 0.0072\, D_2\,.$$

$$(0.5533)^{*}\ (2.9857)^{*}\ (3.5433)\ (0.0769)\ (0.0153)$$

$$R^2 = 0.9362\quad \bar{R}^2 = 0.8773\quad S.E.E = 0.2983\quad D.W = 3.7543$$

The estimates of the regression coefficient have been worked out for three types of holdings. Holding size I for small and marginal holding, Holding size II for medium holding and Holding size III for large holdings. The estimation for each holding has been presented in separate tables. The results of the production function and its elasticity coefficients, t-values and Standard Error have been given in Table 1, 2 and 3. From the above discussion we can conclude that larger size holdings are more economically efficient than small and medium type of holdings because the large type of holdings operate under increasing returns to scale where as others always don't satisfy this criteria. They operate at either Constant Returns to Scale (CRS) or Decreasing Returns to Scale, although many times all farms operates at Increasing Returns to Scale.

Table 3
Estimation of Regression Coefficients for Holding Size – III

Independent Variable.	*Elasticity*	*Standard Error*	*t – value*	*Mean (S.D)*
Intercept	0.9172	1.6492	1.3459	411.75 (51.12)
Fertilizer used (kg/per pond)	0.7985	0.7623	4.3471*	11.925 (4.90)
Seed stocks (No/per crop)	0.6433	1.9755	0.3255**	114.75 (53.32)
Size of tank/pond	0.8732	1.9743	0.4327	134.07 (17.21)
Rate of survival (per 1000)	0.6227	0.7679	2.4562*	107.59 (35.42)
Quantity of feed used (kg/per pond)	0.4983	0.7542	1.7325**	61.37 (44.32)
Power consumption.	0.3255	0.4312	0.5533	87.319 (6.635)
Amount of labour used in hours	0.4117	1.9893	2.9857*	29.57 (14.51)
Level of education/ skill of labour	0.5893	0.4331	3.5433*	32.516 (51.44)
Success rate (D_1:Dummy)	0.0443	0.7753	0.0769	0.5721 (0.123)
Failure or otherwise (D_2: Dummy)	-0.0072	0.6328	- 0. 0153	0.917 0.142

$R^2 = 0.9362$ $\bar{R}^2 = 0.8773$ S. E .E = 0.2983 DW = 3.7543

Source: See Annexure III.

Note: *: Significant at 1 % level
*: Significant at 5% level

CONCLUSION

The main objective of the study is to quantify the effect of education and skill on economic efficiency coupled with technical and allocative efficiency in the aquatic farms. For this purpose attempt was made to estimate the output in a stochastic frontier production function for the cross-section data by using Cobb-Douglas types of production function. The farms are divided into three categories and the study found, larger types of holding are more economically efficient than medium types of holdings. Similarly medium type of holdings are more efficient than small types of holding. Empirically the study shows that there is more disparity in technical efficiency of the farms varying from small types of holding to large types of holding. The variation ranges from 54.75 per cent for small types of holding to 97.43 per cent for large types of holding. Among the sample farms Medium types of holding have both technical and allocative efficiency from productivity point of view. For a given time period and size of the sample the study strongly recommends that the effect of education has more positive impact in increasing farm productivity and subsequently reduce the cost of production. The district agriculture and cooperative department should give more emphasis on launching extension programs to make the people more skilled and adaptive to their working environment.

REFERENCES

Aigner D.J and S. F. Chu (1968), " Estimating the Industry Production Function", *American Economic Review,* 13, 568-598.

Aigner D. L. C. and Schmidt P. (1977), "Formulation and Estimation of Stochastic Frontier Production Function Model", *Journal of Econometrics,* Vol. 16, pp. 21-37.

Bhavani T. A. (1991), Technical Efficiency in India, Modern Small Scale Sector, An Application of Frontier Production Functions." *The Indian Economic Review,* 26, 149-166.

Chaudhari D. P. (1979), "Education Innovation and Agricultural Development: A Study of North India, 1961-72, Vikas Publishers New Delhi.

Farekk M. J. (1957), "The Measurement of Productive Efficiency", *Journal of Royal Statistics,* Series A 120, 253-290.

Fedder G. and R. Slade (1984), "Acquisition of Information and Adoption of New Technology", American *Journal of Agricultural Economics,* Vol. 69-3 August.

Henderson J. M. and R. E. Quandt (1971), Microeconomic Theory, McGraw-Hill, New York.

Kalirajan K. and J. C. Flinn (1981), "Allocative Efficiency and Supply Responses in Irrigated farm Production" *Indian Journal of Agricultural Economics,* Vol. 36 (2) April–June.

Lau L. J. and P. A. Yotopulos (1971), "A Test of Relative Efficiency and Application to India Agriculture, *American Economic Review,* Vol. 61, pp. 94-109.

Lee L. F. (1983), A Test for Distributional Assumptions for the Stochastic Frontier Functions, *Journal of Econometrics*, 22, 245-267.

Moock J. W. (1981), "Education and Technical Efficiency in Small Farm Production, pp. 723-739.

Ramaswamy K.V. (1994), "The Efficiency in Modern Small Scale Firms in Indian Industry Application of Stochastic Production Frontiers", *Journal of Quantitative Economics*,10, 309-324.

Reddy V. Chenna (1976), "Production Efficiency in South Indian Agriculture", *Journal of Far Economics,* Vol. 49 (4), Novermber.

Saini G. R. (1968), "Resource use Efficiency in Indian Agriculture" *Indian Journal of Agricultural Economics,* Vol. 24 (2) April-June.

Schmidt P. (1976), "On the Statistical Estimation of Parametric Frontier Production", *Review of Economics and Statistics,* 58, 238-239.

Sharma R. C. and N. Sinha (1995), "Frontier System and Estimation of Productive Efficiency Intra-Industrial Private and Public Enterprises, *The Journal of Institute of Public Enterprises,* 18, 27-35.

Tilak J. B. J. (1993), "Education and Agricultural Productivity in Asis; A Review", Indian *Journal of Agricultural Economics,* Vol. 48 (2) April.

Waldman D. M. (1984), "A Stationary Point for the Stochastic Fronter Likelihood", *Journal of Econometrics,* 18, 275-279.

Welch F. (1970), "Education in Production Analysis" *Journal of Political Economy,* Vol. 78 (1), pp. 35-39.

13

DO COMMODITY FUTURES GENERATE INFLATIONARY EFFECTS? ANALYZING WHEAT PRICE MOVEMENTS

SANGEET CHAKRAVARTY,
NILABJA GHOSH & SHAILESH KUMAR

INTRODUCTION

A pure theoretical reasoning would suggest that commodity futures trading is a market mechanism with immense elegance. In an economy that is thrown open to the uncertainties of the world demand and supply forces, the futures market is a powerful aid for all players to go on with their business of decision making in a fairly informed manner. Futures market would in principle serve to spread risk over a number of players, stabilize prices and most importantly, would deliver objective information to agents who would otherwise have to rely excessively on personal intuitions and perception. However, in practice the same market has been blamed for at least part of the difficulties that have weighed down on the common man in recent years. The recent spurt of inflation and the spiraling of food prices have been associated by many with the legalization of futures trading in India. Despite, intense debating and the assessments of an expert committee, the matter remains largely unresolved in India. It remains hardly to be said that any institution and its regulatory norms have to be not only efficient and academically appealing, but also politically acceptable for its success in a democratic milieu. In other words, such instruments must not bring desired gains at the cost of welfare losses for large sections of people.

In India agriculture is the livelihood of a large section of the population. Globalization has involved a transition for a protected and inward looking market to an open system that is fraught with uncertainties. The volatilities of the international market driven by the demand and supply forces that emanate from a multitude of countries can bring significant gains to farmers and traders of food grains as much it as they can impose losses that can wipe out their business and even throw many of them into destitution. Such an uncertainty of market could be a serious discouragement to the risk-averse farmers and traders. Grains have a fairly long gestation period, they are moderately durable and can be stored. They are therefore ideal subjects of futures trading. Transactions in these items are normally conducted largely on the basis of medium term expectations on future prices. While such expectations can never be perfect, better availability of information can help to improve the accuracy of estimates. Futures market is an institution, that, through the profit maximization activities of traders and speculators, harnesses the information dynamics of prices in the wider world and more importantly, transmits the effect over to the underlying product market. This function of price discovery is extremely important for guiding the decisions of farmers and traders who are exposed to the uncertain global market. Further, the futures market helps to spread the risk faced by these individual operators over a large body of investors who are willing to buy risk in the financial market across a diversified portfolio. To sum up, the futures market can offer both welfare and efficiency gains in times of globalization.

In practice, the above rationale does not appear to have been adequate for the functioning of the futures market for essential food items in India. The suspension of trading in rice and wheat futures in India after a short experience is an evidence of misgivings that linger around the idea. Futures-trading has been blamed by many in power for raising food prices and for fuelling inflationary pressure that are inimical to large sections of consumers in the country. In fact, even the argument that the market can help to improve the welfare of the farmers is defeated by the possibility of an inflationary impact in India since large sections of the Indian farmer are resource poor, small holders and are often net-buyers of foodgrain. Thus, what is appealing on economic grounds has hardly succeeded in the political economy at large.

This paper aims to revisit the issue of futures trading and explore the role of futures market as a driver of market prices using a time-based GARCH framework. This is done by analyzing the dynamics of commodity prices in the market with three different enquiries (1) did the practice of futures trading make a difference to market prices of wheat? (2) Was the rising wheat price driven by the wheat price in the futures market? (3) Was the market price influenced by the volume of trade in the futures market? We have used daily data on prices and futures market from the official website of National Commodity and Derivative Exchange (NCDEX) in Mumbai.

COMMODITY DERIVATIVES AND WHEAT FUTURES IN INDIA

A Background

A wheat future is a derivative whose value depends on the value of other more basic underlying variables. The futures contract essentially is an agreement to buy or sell an asset at a certain time in future for a certain price. The commodity exchange provides the two trading parties, unknown to each other, a guarantee that the contract will be honoured. The contract is referred by the specified delivery month. The exchange also specifies the amount of asset, the product quality and the delivery location. Several such contracts are traded up to many months in the future. The futures prices are regularly reported in the financial press.

Futures trading actually have a long history in India but the apprehension of an excessive speculative effect emanating from the futures market and impinging on price movements was evident ever since an organized commodity derivative market started in 1875, barely a decade after its emergence in Chicago.

In finance a futures contract is a standardized contract, traded on a future exchange, to buy or sell a certain underlying instrument at a certain date in the future, at a specified price. The future date is called the delivery date or final settlement date. The pre-set price is called the futures price. The price of the underlying asset on the delivery date is called the settlement price.

Sporadic attempts to regulate from the top characterized the progress of the institution, including a ban in 1939 (Pavaskar, 1985[I]). A forward markets commission (FMC) was formed in 1952 to regulate the market but in 1966 the trading was banned altogether. This was the beginning of a period when the government attempted to control prices of many agricultural and essential commodities as a policy. State intervention in the food market continued for decades during which time, the idea of formal futures trading got eclipsed. Only after the reforms were under way was the chapter reopened. However trading in futures had in fact continued informally even when the exchanges went out of business (Sahadevan, 2002). The Kabra (expert) Committee on forward markets, set up by the Government 1990, recommended the strengthening of the FMC and also certain amendments of the existing Forward Contracts (Regulation) Act 1952. Following its recommendations, in 2003 the Govt. of India rescinded all previous notifications which prohibited futures trading in a large number of commodities and set the stage for a revival of futures market in a regulated manner (see Note on FMC). This period went hand in hand with the other measures of market liberalization both in the domestic and global market and an enlarged futures market was proposed for the purpose of hedging the risk and uncertainty that were inevitably associated with the processes of globalisation and for minimizing the wide fluctuations in commodity prices.

The FMC gave recognition to three national exchanges as the first tier of regulation. These are the National Commodity and Derivative Exchange (NCDEX)-Mumbai, the National Board of Trade (NBOT), Multi-Commodity Exchange (MCX)-Mumbai and National-Multi commodity exchange (NMCE) – Ahmedabad. The development of the system and its regulation followed the lines of stock market and SEBI (Economic Survey 2004). Various agricultural commodities were added to the list of traded items including essential food items such as rice and wheat. The NCDEX also pioneered a concept of polling prices from the Mandis/rural regulated markets through authentic agencies and contribute to the statistical data base of the economy. Financial liberalization opens a door to various players including traders, producers, stockers and exchanges to operate, expand and benefit.

Wheat assumes a special place in India's commodity economy because of its significance for the country's food security, trade, industry and agriculture. Till now wheat constitutes more than 30% of India's foodgrain output and accounted for 6% of the trade in agricultural commodities. The effect of the liberalization and globalization of market on wheat prices can hardly go unnoticed. Since the early 1990s when the economy was being thrown open, wheat availability has moved from unmanageable excesses to disquietingly low levels and invariably, the wheat price movement also reflected these swings. The price movement were as much boon to the producers and traders as they were a concern for consumers and the policy maker at times. The legalization of futures[2] In 1980, on the recommendation of the Khusro committee, futures trading on select commodities (cotton, jute, potato etc.) was reintroduced by the Government. Trading in food grains like rice and wheat was undoubtedly a milestone in India's alignment to the global economy.

Wheat was added to the list of traded commodities in 2004 but the trading was disrupted soon in 28th February 2007 at which point the FMC notified that no new wheat and rice contracts will be launched till further notification though running contracts were to continue. The suspension was apparently a part of a broader endeavour to reign in food prices arising from the much familiar misgiving that the practice could encourage unhealthy speculative elements and hoarding activities. Despite several enquiries on the issue, the ban continues till date.

THE PRICE DYNAMICS IN WHEAT MARKET

Price behaviour of wheat is determined by the dynamics of simultaneous buying, stocking and selling of grains depending on speculative expectations. Volatilities arise on account of market uncertainty, indecisions and changes. A conventional measure of volatility is the standard deviation or its scale neutral derivation coefficient of variation (CV), but the validity of these measures decrease in a dynamic setting (Chakravarty, 2007). Autoregrssive moving average (ARMA) models are used for forecasting with uni-variate time series data but when the variance can also be affected by information from the past the series takes a hetero-scedastic from and Engle (1982, 1985) auto-regressive conditional heteroscadasticity (ARCH) models become more suitable. The conditional variance is allowed to change over time depending on the past values of squared errors while the mean and variance are estimated simultaneously for greater effiency (Enders, 1995). A more parsimonious generalized ARCH(GARCH) model was proposed by Bollerslev et.al. (1986). This is akin to an ARMA model in which the conditional variance depends on first p past conditional variances and q past squared unexpected prices. This GARCH (p,q) process is given by.

$$\text{Equation (1) } P_t = \beta_0 + \beta_1 P_{t-1} + \beta_2 X_t + \varepsilon_t \qquad (1)$$

$$ht = \omega + \sum_{i=1}^{q} \alpha_i \varepsilon_{t-i}{}^2 + \sum_{i=1}^{p} \gamma_i h_{t-i} \qquad (2)$$

where P is price of the product in the mean equation (1) and X is any other relevant exogenous variable and in the variance equation (2), h is the conditional variance specified as a linear function of past squared observations and α, β, γ and ω are the parameters to the estimated in the model.

The plot of the wheat prices in figure 1 from the June 13, 2005 to August 20, 2007, shows a non-stationary series (an ADF-test affirm this) with episodic volatilities and suggests that GARCH model could be used to describe the price behavior. The two episodes of high volatility noted in January to April 2006 and January to April 2007 coincide with the events when price rise was arrested by Government intervention and in particular when news of impending imports broke in. Roughly four phases are demarcated as

(1) Up to January 2006........................rising price

(2) January-06 to April –06...................controlled price

(3) April-06 to November –06................rising price

(4) November -06 to April-06................falling price

(5) April-2006 onwards........................Rising price

For about six years, India's food security position was comfortable and India even emerged as a wheat exporter in the global market. When the prices start rising, imports were contemplated after a long time. A few tenders were floated in the global market in the second phase in which time the price rise seemed to have been arrested. The third phase was a reversal to the rising price situation in which further tenders were floated and imports made. With about 3 million tones of import having arrived by October, the price rise was arrested again from November. The period was also marked by news of a possible global shortage (and of poor harvests in countries like Brazil and China coming from USDA). Repressive measures like a ban on whet exports were enforced even in a post-liberalized policy regime. In February, the government took further measures to control prices that included the ban on futures trading. In March there were further tenders for imports which however were to be cancelled though imports could not be averted subsequently. The period of rising prices was there to stay even after the ban.

For modeling price behaviour through a GARCH model the data needed to be corrected for the effect of oil price movements and also for its regular seasonal effects. A regression of an ordinary least squares (OLS) type was fitted using the price of diesel as an independent variable and a dummy for the three-post harvest months April, May and June during which price could be expected to dip. While diesel price was found to have a highly significant coefficient the season effect, though expectedly negative, was insignificant (Table 1). The residuals were tested for ARCH effect with a positive result Appendix (Table A1(a)) and heteroscedasticity could not be ruled out. Based on usual diagnostics (AIC, SBC, RBAR-sq.), the GARCH(1,1)equation was finally estimated. The GARCH(1,1) equation shows expected results with positive and stable effects and the sum of the effects (0.285 and 0.722) is close to one suggesting that volatility is not persistent Appendix A1(b).

Table 1
Estimation of OLS and GARCH Equations on Wheat price

Dep=Log Price	*OLS*		*GARCH(1,1)*	
	Coeff	*T-Statistic*	*Coeff*	*Z-Statistic*
Log(price(-1))	0.984	186.79	0.995	395.91
Log(Diesel price)	0.035	2.42	0.017	7.126
Dummy Harvest	-0.001	-0.927	-0.0004	-0.624
Constant	-0.107	-1.57	-0.068	-6.912
Adj.R Sq	0.99		0.99	
F-Statistic	25514		126	

We then ask if the price dynamics of wheat was possibly impacted by the introduction of the futures market by using a dummy to demarcate the period in which the market existed. Introduction of the dummy variable for post-futures period does not rule out the influence of futures trading on price movements. The coefficient is negative but insignificant when the effect on mean equation is examined but the positive coefficient in the variance equation suggests that the futures market operation may have brought down the price movements or volatilities in the wheat market.

CO-MOVEMENT OF FUTURES AND SPOT PRICES

Determination of derivative prices especially in the context of equities have been modeled and analyzed in various research

Table 2
Estimation GARCH(1,1) Equations on Wheat Price with Dummy for Futures Trading Years

Dep=Log Price	*Model 1*		*Model 2*	
Variables	*Coeff*	*Z-Statistic*	*Coeff*	*Z-Statistic*
Constant				
Log (price(-1))	0.9966	428.249	0.9937	287.7037
Log (Diesel price)	0.0100	3.5286	0.0162	2.3145
Dummy Harvest	-0.000	-0.0906	-0.0005	-0.5976
Dummy –futures trading	-0.0010	-1.1811		
Variance equation				
Constant	2.31E-06	3.3837	7.89E-06	6.0932
ARCH (1)	0.2699	10.6106	0.2631	11.833
GARCH (1)	0.76603	41.4387	0.6676	23.2596
Dummy –futures trading			5.70E-06	1.7094

Note: See table A1 (b) for GARCH effects.

works (Hull, 1997, Sengupta 2005). The futures price is a valuable signal, conveying additional information to agents in both markets so that they interact mutually and move together contributing to the whole price discovery process. The information transmission between the derivative and the underlying market have also drawn attention in literature (Allophones, 2002, Tse, 1999 Kenourgios, 2004), that generally, found indications of bi-directional information flows. In fact it is generally believed that the futures market assimilates information faster than the underlying product market (Bose, 2007) and could play a 'leading' role in the system. In this sense, futures market can help to make the underlying product market also efficient through information transmission.

To examine the transmission process, the correspondence between wheat market price and futures price were examined using a conventional tool. Futures contracts for wheat in India were traded separately and often concurrently with different expiry dates. We have considered eight such contracts within our sample period corresponding to expiry dates October 2005, January 2006, April 2006, July 2006, October 2006, January 2007,

April 2007 and July 2007 spread over the periods occurring both before and after harvest. The last two contracts (distinguished by the expiry dates) April 2007 and July 2007 provides partial and diminished pictures of trading only. Each contract is traded at varying prices till expiry. The futures prices would differ from current spot prices depending on short term and long-term factors and interest and storage costs, traders expectations processes an the regulatory norms but in on ideal situation the prices would be close and would move together with no particular bias.

A visual inspection of Figure 1 does not affirm a co-movement. The spot price appears rigid in its downward adjustments relative to the futures price. In fact the multiple contracts do not uniformly show similar directions of movement (such as the January 2006 contract reaching a trough while the April 2006 contract shows a steady rise).

Figure 1: Current and Future Prices

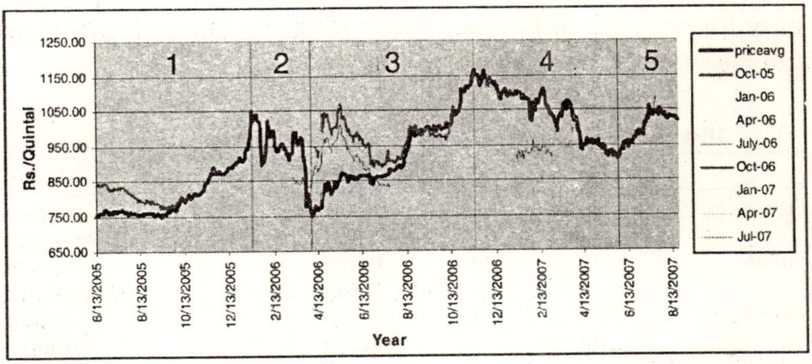

No definite conclusion emerges on the directional correspondence of movements of the two prices from the correlation analysis either. The only two contract periods for which a reasonably high correlation is noted are both January in the years 2006 and 2007. The correlation coefficients are in general low and sometimes even negative. During the two volatile periods the correlation coefficients are found to be negative (Table 3).

Table 3
Cotemporaneous Correlation Coefficients between Wheat Prices and Wheat Future Prices

Contract Period	*Future Contracts*	*Correlations*
10 June 05 to 20 Oct 05	Oct-05	0.01
10 Sept 05 to 20 Jan 06	Jan-06	0.56
10 Dec 05 to 20 April 06	Apr-06	-0.34
10 Mar 06 to 20 July 06	Jul-06	-0.18
10 Apr 06 to 17 Oct 06	Oct-06	0.28
10 Aug 06 to 19 Jan 07	Jan-07	0.77
10 Oct 06 to 20 Apr 07	Apr-07	-0.35
10 Jan 07 to 20 July 07	Jul-07	-0.36
Unstable periods	*Apr*	*Jul*
Jan 06 to Apr 06*	-0.51	-0.32
Jan 07 to Apr 07**	-0.60	-0.25

- *Correlation with Apr/July 2006 prices·
- ** Correlation with Apr/July 2007 prices

Testing for co-integration, there appears no long run relation binding the two prices for any of the contracts. The hypothesis of no binding relation could be rejected at 1% level only for the case of January 2006 (Table 4). The two prices do not show a tendency to move together[1].

Table 4
Test for Co-integration between Future and Current Price

	Eigenvalue	*Likelihood Ratio*	*Hypothesized No. of* CE(s)
Oct-05	0.106	13.851	None
	0.030	2.979	At most 1
Jan-06	0.243	32.195	None **
	0.034	3.511	At most 1
Apr-06	0.124	17.550	None
	0.031	3.369	At most 1
July-06	0.082	13.445	None
	0.040	4.312	At most 1
Oct-06	0.088	16.864	None
	0.018	2.787	At most 1
Jan-07	0.076	13.625	None
	0.027	3.461	At most 1
Apr-07	0.045	8.400	None
	0.009	1.428	At most 1
July-07	0.053	1.907	None
	0.004	0.121	At most 1

Note: 5 % Critical Value is 19.96 (None) and 9.24 (At most 1). 1 % Critical Value is 24.6 (None) and 12.97 (At most 1)

VOLUMES TRADED AND THE PROBABLE ROLE OF INSIDER INFORMATION

The relation between futures prices and trading volumes has been of interest to many scholars studying the behaviour of stock markets. (Soydemir and Petrie, 2003, floros and Vogues, 2007). Although we have found no positive relation between futures price and the market price of wheat, it is however possible that a direct relation between trading volume in the futures market and wheat prices may exist without the intermediation of futures price as a transmission medium and other implicit media can work towards this transmission. In fact the trading volume may incorporate relevant information on the nature of wheat market. One appealing rationale for this hypothesis is that the volume of transactions in futures market could be a measure of the speed of market evolution in which case volume would have a relation with the price change variance (Clark 1973). In such a case, the trading volume can be used as an instrument for the true operational time measuring the speed of evolution of the price process. There could however be two possibilities with different implications given that the arrival of new information in the market leads to price volatilizes. When new information arrives sequentially or when insider information is available to certain parties first, expectations and responses could be diverse among the players (some traders adjust their expectations up and others down) leading to active trading. As a consequence, large price movements would be associated with high trading volumes. On the other hand, if the information penetrates the market completely, expectations and responses are likely to be in similar directions so that actual trading volumes would be low despite price changes. In the literature on financial markets the two variables of interest are found to move usually in the same direction. In Clarke's own example there is a positive correlation between volume and the magnitudes of price movements.

In this part of the analysis we consider only data from the period during which there was active trading in the form of open interest declaration. This starts from June 13, 2005 and runs up to April 20, 2007. We first correct the prices for the effects of oil price changes and seasonality through an OLS model using a logarithmic specification as before[1]. Using the corrected prices we construct our study variable DLP2 measuring the fist difference or the price change variable so that

$$DLP2= \Delta \text{ Log(PRICE2) with}$$

Log(PRICE2) being the logarithm of wheat market price corrected for the effects of seasonality and oil price movements.

We begin by enquiring about the behavior of the variable DLP2. Under a hypothesis of market efficiency, trading would take place in a way so that current or past price changes would convey no information on future price changes (Clarke, 1973). In such a case we would expect price changes to be stationary with an expected value of zero. An ADF test reveals that the series is stationary. However normality is not indicated. The median value is near zero consistent with efficiency but the variable is negatively skewed so that the mean is negative. This means that larger changes are more frequent than smaller changes. The values are highly clustered leading to a typical fat tail distribution.

Table 4
Test Statistics for Study Variable of Price Change DLP2

ADF Test Statistics	*-11.2 (1% Critical Value=-3.4)*
Mean	-0.00001
Median	0.00008
Skewness	-4.8
Kurtosois	50.1
Probability (Jarque-Bera normality test)	0.000

We specify the volume variable by

DVOL = ΔLog (Volume traded) and estimate GARCH equations for the study variable.

We first include the volume in the mean equation and find a negative effect of trading volume on price though the coefficient is not significant at 1% level. In any case a positive role of trade in futures market on wheat price rises is not confirmed. Second we include the volume in the variance equation to examine its effect on the magnitude of price changes. The effect of volume being considered as contemporaneous can however create simultaneity problem since it is likely that the magnitude of price changes also feed into the information instantaneously to determine the volume of trade. In general price instability is considered favorable for a trader's activities. To take care of this possibility we have also used an instrumental variable method, using past values of volume as instruments. In model 2, the estimated coefficients of the lagged dependent variable, as well the GARCH and the ARCH effects are all highly significant but in the model3 involving an instrumental variable only the ARCH effect is significant. The GARCH effect of volume is found to be negative and highly significant in both cases i.e, with and without instrumental variable method, suggesting that low volumes are consistent with high volatilities in product market. This might imply that when information is scarce, it reaches the market sequentially and market evolution is slow. When information keeps arriving in a stream it penetrates the market completely as in the situation when the news of import tenders are published and trading volume is low.

Table 5
Effect of Futures Trading Volume on Wheat Price Movements through GARCH(1.1) Models

	Model M1		*Model M2*		*Model M3 (Instrumented)*	
Dep=DLP2						
	Coeff	*Zstatistic*	*Coeff*	*Zstatistic*	*Coeff*	*Zstatistic*
Mean Equation						
C	0.00032	0.327	.00014	0.151	0.00060	0.7520
DLP2 (-1)	0.1211	1.691	0.1700	2.156	0.0738	1.0537
DVOL	-0.0014	-1.729				
Variance Equation						
Constant	0.0002	17.404	0.00014	8.489	0.00020	31.355
ARCH (1)	0.2232	5.860*	0.1530	6.181*	0.3360	6.592*
GARCH (1)	-0.0377	-0.628	0.3912	6.232*	0.0138	0.594
DVOL			-0.0001	-12.62*	-0.00009	-22.914*

Note: * Significant at 5%.

CONCLUDING REMARKS

The opening of the futures market for agricultural commodities is undoubtedly a milestone in the progress of financial reforms in India. In principle a futures market can help commodity markets in discovering the efficient price by harnessing added information from the larger market. The timing of the futures trading activity in wheat market in India is significant as it allows us to understand the implications at a relatively insecure moment for the country. Wheat is an important commodity in India's political economy not only for its role in food security but also for the significance it holds for the farm sector, the processing industry, the country's international trade and above all for public food policy. Moreover, wheat came to be a commodity to be traded in the futures market in the early 2000s. The period of this study starting from June 2005 was marked by the usual reform process in the market, a specter of global shortage and a tendency of rising price in the domestic and global market.

The Graphical plot of the market prices showed a non-stationary behavior and revealed that pronounced price volatility was coincidental with the government intervention at controlling the price and with news of impending import that the media published. The weak correspondence between the spot and futures prices however rules out the role of future prices in dragging the market prices upwards even though that would indicate probably that the futures price served as a mere reflection of the international and domestic factors that are relevant. The lack of correspondence also puts a question mark on the efficient functioning of the market. Interestingly, when considering trading volumes in the futures market, a significant relation with price movements was noted. The result of the analysis also raises a possibility that when market information is scarce, the same information may actually be exclusive to a select few. On the whole something in the way of futures market functions tended to obstruct the free flow of information between the prices, defeating the purpose of the futures market. Worse, while trading volumes conveyed information from the futures to the product mark, this does not make up for the inefficiency of functioning and the nature of the information dissemination may neither by symmetric nor be consistent with the welfare goals of the economy.

Appendix

Table A1 (a)

ARCH TEST of Residuals of OLS in Equation 2

p	*F-Statistic*	*Observed* R-Squared*
1	19.28	18.79
2	9.85	19.21
3	7.12	20.81
4	5.32	20.77

Table A1 (b)

GARCH TEST of Equation 2 with Alternative Specifications

(p,q)	*Const.*	*ARCH(1)*	*ARCH(2)*	*GARCH(1)*	*AIC*	*SBC*
		$\alpha 1$	$\alpha 2$	$\gamma 1$		
(0,1)	5.57E-05	0.685			-6.451	-6.410
(0,2)	4.46E-05	0.643	0.174		-6.489	-6.442
(1,1)	4.38E-06	0.285		0.722	-6.590	-6.543
(1,2)	4.34E-06	0.316	0.018	0.695	-6.586	-6.532

Notes: The forward markets commission (FMC) is a statutory body that functions under he administrative control of the Ministry of Civil Supplies consumer affairs and public distribution, Govt. of INDIA. It is entrusted with the task of advising the govt. keeping forward markets under observation. The instruments for regulation employed by the Commission include ordinary a special margins, limits, maximum and minimum prices, security deposits, suspension of trading and prohibition. When forward contracts are prohibited or suspended in certain commodities including an essential commodity like foodgrain indulgence in such contracts is an offense attracting penal provision.

NOTE

1. Further tests such as the Granger causality test also could not confirm the leading role of the futures price is driving the market price (For more details see Ghosh, Chakravarty and Kumar, 2007).

REFERENCES

Alphonse Pascal (200), Efficient Price Discovery in Stock Index: Cash and Future Market, *Journal of Futures Markets*, 19,911-30.

Bollerslew, T. (1986), Generalized Autoregressive Conditional Hereroscedasticity, *Journal of Econometrics*, 31, pp. 307-327.

Chakravarty, Sangeeta (2007), Changing Pattern of Stock Volatility in India, *The Indian Journals of Economics*, Part-4, Vol. LXXXVII

Chand Ramesh, S., Raju and L.M. Pande (2007), Growth Crisis in Agriculture, Security and Options at National and State Levels, *Economic and Politically*, June 30.

Clark Peter K. (1973), A Subordinated Stochastic Process Model with Finite Variance for Speculative Prices, *Econometric*, Vol. 41, No. 1, January.

Engles R. F. (1982), Autoregressive Conditional Heteroscedasticity with Estimates of U.K. Inflation, Econometric, 50, Pp. 987-1008.

Engles R. F. (1995), Applied Econometric Time Series, John Wiley

Floros Christos and Dimities Vogues (2007), Trading Volume ad Retunes Relationship in Greek Stock Index Futures Market: GARCH vs GMM Intenational Research Journal of Finance and Economics, 12.

Ghosh Nilabja, Sangeeta Chakravarty and Shailesh Kumar (2007), "Volatility and Price Discovery in Indian Wheat Market: Was the Futures Market to Blame?" Paper presented in the seminar on " Agri-Commodity futures in India: Impact on Production and Prices" Organized by MCX-AER and ISAE at University of Mumbai on Sept.28,2007)

Hull, J. C. (1997), Options, Futures and Other Derivatives, Prentice Hall International Inc.

Kenourgios Dimities F. (2004), Price Discovery in the Athens Derivative EXCHANGE Economic and Business Review, Vol. 6, No. 3.

Pavaskar Madhoo (1985), Saga of he Cotton Exchange, Bombay Popular Prakshan.

Sahadevan K. G. (2002), Sagging Agricultural Commodity Exchanges in, *Economic and Political Weekly*, July 27.

Sengupta Ambar N. (2005), Pricing Derivatives: The Financial Concepts Underlying the Mathematics of Pricing Derivatives, Mc Graw-Hill.

Soydemiar Gokce A. and A. George Petrie (2003), Intrude Information Transmission between DJIA Spit and Futures Markets, *Applied Financial Economics*, 13, 817-827.

14

AN EVALUATION OF DETERMINANT AND BENEFITS OF CO-OPERATIVE LED CONTRACT FARMING: ACASE STUDY OF SUGARCANE CULTIVATION IN ORISSA

B. B. Swain

INTRODUCTION

In recent debate on agricultural development, efforts from every part of the economy have been made to improve the agricultural performance through two instruments; technological innovation and institutional innovation. One institutional innovation is being prompted in recent for agricultural development is contract farming. In this system, a central processing or exporting unit purchases the harvest of independent farmers. Contract farming envisages a symbiotic relationship between agriculture and industry. In definitional term it refers to an alternative market institution, which establishes an agreement (formal or informal) between grower(s) and firm(s) (Exporters, processors, retail outlets, or shippers, for example) to produce and supply the agricultural commodity under forward contract. This forward contract includes the price mechanism, production strategy and supply of inputs. It allows all producers to reduce the transaction cost of selling a perishable product in uncertain or thin markets, and to get higher prices from a buyer who is certain that the farmer will deliver the commodity. In this system the farmer can cultivate his own land and without losing his land title. It is an agreement between farmers and agribusiness firm (or private capital) for the production and supply of agricultural products under forward agreements frequently at pre-determined price. In this process, the company provides farmers with credit, extension services and other inputs; sometimes it provides the crop collection facilities, repaid through deductions from sales to the processing plants, sometimes. It allocates risk between producer and contractor. Risk in agriculture is defined as R, where R = f (Q, P, I).

Here Q is output, P is price and I is input.

In contract farming system the industry handle the price risk which is generated by market and input risk. Where as the farmer handle the production risk. In this system the role of State in agricultural marketing and input supply is being reduced and an increasing space is provided for private sector to bring better market efficiency in in-put and out put markets.

Minot (1986) R. Mighel and L. Jones as originators, in 1963, have widely used classification of contracts into three kinds, namely "market specification", "resource providing," and "production management". In the first modality, the transaction between growers and buyers is agreed in terms what to be produced and what are the commitment for future selling commodities. The second modality tells what inputs should be provided and in which way to farmers and lastly third modality tells how to produces and what are the procedures. According to Eaton *et al*, (2001), the nature of contractual arrangement takes place in one of the following forms depending upon the structure of the economy, i.e. (a) The Centralized Model, (b) The Nucleus Estate Model, (c) The Multipartite Model, (d) The Informal Model and (e) The Intermediary Model.

OBJECTIVES

The main objectives of this study are as follows.

(1) To evaluate the logic of contract farming system

(2) To evaluate the benefits of sugarcane cultivation through contract farming system

(3) To examine the determinants to participate in this system

This paper first discusses some aspect of the notions of 'importance of contract farming' and 'the experience of contract farming system' and their realities. More recently, some economist and policy makers are emphasizing that the contract farming system is suitable for Indian agriculture whereas others are going against this. Finally the paper will turn to the field experience of contract farming system that the performance of contract-actual arrangement between farmer and agriculture and what the constraints for contract system in this area are. The last section reflects the summary and conclusion.

DATA AND METHODOLOGY

The present paper is based on primary data. For collection of primary data, the study has taken one village (Hatisara) from Orissa purposely; because this village economy is closely associated with the sugar industry, where the farmers are cultivating the sugarcane and supplying the sugarcane to a sugar company on a contract basis. The statistical information concerning the village has come from a detail study of all the 141 households with an emphasis to analyses the information about Sugar Company and sugarcane cultivation and resource position of these households. The survey done by personal interview on the basis of a suitable questionnaire. The information was collected with respect to the composition and occupational structure and their assets and their credit situation. Apart from the questionnaire, qualitative information has also been collected about the nature of contractual arrangements between farmers and the sugar industry.

To find out the basic needs for sugarcane cultivation this study uses Logit model by taking farmers who are cultivating sugar cane and who are not cultivating. Logit model extends the principles of generalized linear models (ex. regression) to treat better the case of dichotomous and polychromous dependent variables. The Logit model gets its name from the "logistic" functional form of the choice probabilities, which traces out an S-shaped curve. Logit regression model applies maximum likelihood estimation after transferring the dependent into a Logit variable (the natural log of the odds of the dependent occurring or not). The Logit probability of a certain event i occurring can be expressed as;

$$P_i = E(Y = 1/X_i) = \beta_1 + \beta_2 X_i \tag{1}$$

Where Y = 1 means the farmer cultivating the sugarcane and X is total land owned by a farmer. But the probability of cultivating the sugarcane can be represented as;

$$P_i = E(Y = 1/X_i) = \frac{1}{1 + e^{-(\beta_1 + \beta_2 X_i)}} \tag{2}$$

For ease of exposition, we write (4.2) as

$$P_i = \frac{1}{1 + e^{-Z_i}} = \frac{e^z}{1 + e^z} \tag{3}$$

Where $Z_i = \beta_1 + \beta_2 X_i$

If P_i, the probability of cultivating the sugarcane, is given by (4.3), then (1- P_i), the probability of not cultivating the sugarcane

$$1 - P_i = \frac{1}{1 + e^{Z_i}} \tag{4}$$

Therefore, we can write

$$\frac{P_i}{1 - P_i} = \frac{1 + e^{Z_i}}{1 + e^{-Z_i}} = e^{Z_i} \tag{5}$$

Now $P_i \backslash (1 - P_i)$ is simply the odds ratio in favor of cultivating sugarcane-the ratio of the probability that a farmer will cultivate sugarcane to the probability that it will not cultivate sugarcane. Now if we take the natural log of (4.5), we obtain an equation known as Logit model.

$$L_i = \ln\left(\frac{p_i}{1-P_i}\right) = Z_i = \beta_1 + \beta_2 X_i + U_i \tag{6}$$

Here L is called the Logit and hence the name Logit model.

In the present study we consider four independent variables for analyzing the Logit model;

$$L_i = \ln\left(\frac{P_i}{1-P_i}\right) = \beta_1 + \beta_2 TL + \beta_3 IL + \beta_4 CR + \beta_5 INST + u_i \tag{7}$$

Here

TL = Total land of a farmer/households

IL = Total irrigated land of farmer/households

CR = Total credit taken by a farmer/households for cultivation purpose

INST = Total number of instruments (plough, pump sets, or cart) of a farmer/households

IMPORTANCE OF CONTRACT FARMING

Risk and uncertainty[1] that results from it are not new to agriculture. The farmers and ranchers make decisions in a risky environment every day. The consequences of their decisions are generally not known and the outcome may be better or worse than expected, when the decisions are made. Bharat Ramswami et al, (2003) mention that there are three types of risks (production, price (or market) and input) that exist in agriculture, which the farmer faces in their daily life. Production and price are two major risks in agriculture that confront farmers. Technology changes, legal and social concerns, and the human factors itself also contribute to the risk environment for agriculture producers. So to avoid or to reduce the risk in agriculture there is needed an alternative institutional arrangement in this risk environment. This requires an evolution of tradeoff between change in risk, expected returns and entrepreneurial freedom among others. Contract farming is alternative risk management system, which reduces the risk in agriculture by sharing between farmer and industries, Pari Bauman (2000), Eaton et al (2001), World Bank (1989).

At Macro level contract farming can help remove market imperfections in produce, capital (credit), land, labour, information and insurance markets facilitate better co-ordination of local production activities which often involve initial investment in processing, extension etc,; and can help in reducing transaction costs (Grosh, 1994; Key and Rusten, 1999). It helps for crop diversification and for improving farm incomes and employment (Benziger 1996; Singh, 2000). Contracting gives access to additional sources of capital and more certain prices by shifting part of risk of adverse price movement to the buyer (Hill and Ingersent, 1987). It helps small holders to produce primary production (annual or tree crops, live stocks diary, poultry, eggs, fish shrimps, etc) and make it superior 'nucleus enterprises, which would handle one or move more of the upstream and down stream activities such as input supply, output processing and marketing by making it an institution (white, 1997). The firm even gets an access to unpaid family labour (white, 1997) and can make use of state funds indirectly through agricultural production sector, which are directed at farmers by development agencies (Clapp, 1988).

In the context of economic reforms there are several reasons identified in the literature for the introduction of contract farming as a method of organization of production in agriculture in developing countries like India.

The structure of the Indian agriculture is dominated (and will continue be dominated) by smallholders less than two hectares of land. The dominance of small farmers leads to major production of low value crops that leads to low level of income earned by farmers. These small farmers have traditional cultivation, management and seasonal production practice and are independent producers. They sell their produce individually, have little bargaining power with input suppliers and produce market. On the other side, there has been a changing food habit of people because of increase in the income of the middle class people due to urbanization and modernization. It is a challenge to small farmers to increase their income by shifting crop cultivation from low value to high value through adoption of new techniques of production, (Gulati, 2006). To access the upcoming urban and international market, contract farming is seen as one option wherein the farmers produce the agricultural products marketing firms (Banerjee *et. al.* 2002, Gulati 2006, and Dev *et al* 2005).

The sluggishness of growth rate of productivity in agriculture is due to low investment by public sector in agriculture. Contract farming system can encourage private sector to invest more in agriculture (World Bank, 2003)[2]. The implicit opinion is the existence of technology appropriate to Indian agriculture but the technology need to be transferred to the farmers. The sponsoring firm can facilitate in the transfer of technology by having a contract with the farmers to purchase the output produced by the farmers at pre-defined prices. In the process, farmers can get access to production services and credit as well as knowledge of new technology, (Eaton et al 2001, Kumar 2oo6, Banerjee *et al* 2002).

In less developed economy small farmers have difficulty in participating in the market economy, as forward and backward linkages are weak. Contract farming is a substitute to the market system under conditions of incompletely formed markets by providing these forward and backward linkages to farmers as to appropriate the gains from trade. In the process, contract farming so can provide cost efficient inputs, such as extension advice, mechanized services, seeds, fertilizer and credit and guaranteed and profitable markets for their output, (Eaton 2001, Singh 2000, 2002, 2005,).

One of the most important points is that due to lack of efficient infrastructure management and bottlenecks at market level. To avoid these losses and add benefits to the small and marginal farmers a massive improvement is needed in the existing marketing, which can be done by contract farming system because the private (corporate) sector and MNCs would tackle these problem in their own interest and offer better price to farmer, (Gill 2004, Dev *et al* 2004, Eaton 2001).

THE EXPERIENCE OF CONTRACT FARMING: SPECIAL REFERENCE TO INDIA

The incidence of contract farming has remained a controversial subject in the literature. The business proponents advocate contract farming as suitable for agricultural development. It gives a boost to increase the income of farmers and transfers the modern techniques of production to rural areas. Survey by Goldsmith (1985), has a significant positive impact on rural income in almost every case, and contract farming in these cases led to increased productivity and increased cash sales. Studies by Pari Bauman (2000), found that in Kenya, the contract farming system through Kenya Tea development Authority (KTDA) has been a success because of the states support for the scheme. The opponenets of this system give a counter argument regarding the benefit of the small farmers. They argue that contract farming rejects the idea of equal partnership in these schemes. It may subjugate the peasantry to increased control and exploitation by capital, leading to a peculiar from of prolaterionization (Fede 1997; Payer 1980). A section of critic argues that the organization of production in contract farming is one mode of capitalist penetration of agriculture for capital accumulation and exploitation of the farming sector. This even leads to processes of self-exploitation of the farmers and the companies gain indirect control of land. Michal Watt has been a particular outstanding critic of contract farming. Drawing on studies from North America, Africa and Asia, he observes that the history of contracting is replete with company manipulation and abrogation of contracts and the basis of peasant contracting is essentially self-exploitation, labouring more intensively and extensively using children and other non-paid household labourers (Watts, 1994 PP. 63-64).

Like different countries of the world, India has a unique experience on contract farming system. The contract farming system was going on in the colonial period when the British ruled India. However, launching its agro-business in India with special focus on exports of value added processed foods; Pepsi Foods Ltd. entered India in 1989 by installing a tomato processing plant at Zahura in Harshiapur district of Punjab.

PepsiCo followed the contract farming method, whereby the cultivator plants the company's crops on his own land and the company provided selected inputs such as seeds/sapling, agricultural practices, and regular inspection of the crop and advisory services on crop management. Subsequently PepsiCo and other companies, (Nijjar Agro Food Ltd, Hindustan Lever Ltd, and HLL) have used similar methods for the cultivation of food grains (basmati rice), spices (chilies) and oil seeds (groundnut) as well, apart from other vegetable crops such as potato[3]. In Tamil Nadu, Appachi Cotton Company (ACC) has undertaken contract cotton farming with eight farmers groups from 32 villages in Coimbatore district for bringing 1050 acres under cotton farming. The contract growers form an association of persons (AOP). The major features of the model adopted by ACC are: one village-one SHG, one village-one variety of cotton, cotton crop insurance, door delivery of agricultural inputs, crops loans at 12 per cent rate of interest, farm service centers, assured buy back from farmers through farmers are free to sell else where if they find price is higher than contract price. Most of the studies found that contract production gave much higher (almost three times) gross returns compared with that from traditional crops of wheat and paddy. Some studies found that in case of potato in some region of India, producers end up making net losses and traders make substantial profits from the same crop (Mitra and Sarkar, 2003). At present, the contract farming system is going on in every corner of India. Some of the prominent examples would be: mushrooms production in Haryana, Sunflower in Andhra Pradesh and Karnataka, Gherkin production in Karnataka and Andhra Pradesh, Fruits and vegetables production in Tamil Nadu, Maharastra and Andhra Pradesh etc.

RESULTS AND DISCUSSION

Mode of Operation of Contractual System in Study Area

My experience of contract production is based on the fieldwork in one-study location, at the Hatisara Village in Orissa. Present study is concerned with small sugarcane growers who are operated by Bijayananda Cooperative Sugar Mill Limited[4]. This industry commenced in 2004 after closing down of the Ponni Sugar Industries[5]. The selection of farmers by sugar industry is not based on the size of holdings, field history, economic condition of the farmers, source of irrigation and his level of commitment to the contract. That means there is no specific condition for entering into the contract system. If a farmer wants to cultivate sugarcane and supply it to the industry then the farmer informs this to the field officer appointed by the industry. After that, the industry issues a card in the name of the farmer that this farmer is a sugarcane grower of the industry.

To get maximum yield of cane, hybrid seedlings are given to the contract growers by the industry, if the farmer wants. A set of farmer's implements consisting pump sets, credit, other inputs needed in the processes of sugarcane cultivation are also given to the contract grower. The industry was deducting the cost of these inputs in the time of payment to farmer in respect to supply of cane by farmers to industry. The improved technology of cultivation is being channalised to the farmers with the extension support in the form of regular field visits by the technical staff (filed executives) of the industry. The farmers in their own land grow the sugarcane, by self-employing labor. The company fixes the procurement price of cane by working out the cost of cultivation and a minimum profit at average yield and negotiation between farmer and industry. The profitability is compared with its competitive crops like paddy and other vegetables also. The contract farmers in the study area had to transport their produce to the sugar industry at their own cost, sometimes the company provides the transport facilities to the farmers. The payment is made to their respective accounts in the Regional Bank within 10-15 days of procurement.

In this village the predominant crop is paddy (77.47% of the land) and second is sugarcane. The irrigation facilities are not good, only 30 per cent of lands are irrigated through the minor irrigation and 70 per cent of lands are rain-fed but there is no major irrigation facility. Out of 141 households, 31 households are cultivating sugarcane out of 81 land owned households. This village has a low turnover in all markets in respect to land, credit, output and factor inputs.. In this type of economy the Bijayananda Cooperative Sugar Mill Limited tries to change the cropping pattern from low value crop (Paddy) to high value crop (Sugarcane) by providing some incentives to farmers.

Characteristics of Cultivators and Non-cultivators of Sugarcane

The economic condition/resource position contains land, irrigated land, credit taken and instruments (plough, pump sets, and cart). The study village has total 141 households, where 81 households have land. Out of 81 households, 31 households are cultivating sugarcane. The total land of this study village is 238.9 acres, out of this amount of land, sugarcane cultivators holds 54.16 percentage of land, and non-cultivators of sugarcane hold 45.84 percentage of land. If one sees the average land owned by households, the sugarcane cultivators hold more land (4.17 acres) than the non-cultivator of sugarcane (2.19 acres). In case of irrigated land, the sugarcane cultivators holds more irrigated land (76.60 per cent) of total irrigated land than the non-cultivator sugarcane (23.40 per cent). Average irrigated land-holds by sugarcane cultivators (1.79 acres) is higher than the non-cultivators of sugarcane (0.34 acres). Like wise credit taken by the sugarcane cultivators is higher than the non-cultivators of sugarcane and also the average credit taken. In case of instrument distribution the cane cultivators contain more instruments than the non-cultivators of sugarcane. The above analysis shows that the economic condition of sugarcane cultivators is better than the economic condition of non-sugarcane cultivators. So it can be said that by entering the contractual arrangement, the farmer's conditions can change from low level to high level.

Economics of Sugarcane Cultivation

The gross returns from different crops production are presented in Table 1. Total variable cost constitute the costs of human labour, bullock labour, machine power, seed, irrigation and interest on working capital, which the present study did not take for analysis because of lack of information. The study has taken only the output level of different crops. Per acre yield and gross return of sugarcane cultivation is higher than the other crops. If one compares between the sugarcane and non-sugarcane cultivators, sugarcane cultivators are getting higher gross return than non-sugarcane cultivator in every crop except black gram. This shows that the sugarcane cultivators are more efficient in cultivation practice than non-sugarcane cultivators.

Table 1
The Economic Condition of Cultivators and Non-Cultivator of Sugarcane

Indicators	*Crops*	*Area*	*Output*	*Output per Acre*	*Amount per Acre*
Sugarcane Cultivator	Sugarcane	20.45	6490	317.36	300,542
	Rice	66.3	646	9.74	4188.2
	Sunflower	5	5.22	1.04	5200
	Green Gram	6.15	3.8	0.62	1240
	Black Gram	3.75	2.05	0.55	1100
Non-Sugarcane Cultivator	Rice	119.5	850.5	7.12	3061.6
	Sunflower	2.65	2.23	0.84	4207.547
	Green Gram	4.35	2.6	0.60	1195.402
	Black Gram	1.05	1.2	1.14	2280

Source: Field Survey,
Note: Area in acre, Output in quintal, Amount in Rupees.

Determinants of Participating in Contract Farming of Study Area

What small farmers seem to want from contract schemes is essentially a satisfactory regular cash income, and in some cases, additionally, the availability of inputs normally unavailable or more expensively obtained through other sources (notably credit facilities and fertilizer). Their achievements, however, are governed by and embedded in a broader socioeconomic context, which encompasses and reflects the power relationship between small producers and big business men. Small farmers need to know not only that they gain financially from the contract enterprise, but also that they are not being cheated out of further profits by company maneuvering (Gina Porter and Kevin Phillips-Howard, 1997). Freely entered into the contract allows growers to make better use of their specific endowments in imperfect markets and to arrive at combinations of income, effort and risk reflecting their resources and tastes (Little and Watts, 1994). The relationship between the firm and the farmer in a contract would be smooth when both are locked in specific investments. As survival of both parties is closely linked the contract is self-enforcing. The distribution of this advantage (like share the risk between farmer and company) depends heavily on factors like bargaining power, availability of alternative information. The advantages to a party in any one of these factors ex-post, that is after investment has led to its opportunistic behaviour which is defined as unanticipated non-fulfillment of contract (Williamson, 1986). This leads to strain the contractual relationship and may result in failure of the contract. The present study analyses what are the basic inputs that are needed for farmers to shift from food crop to non-food crop (sugarcane). There are 31 farmers cultivating sugarcane in this village and are supplying the cane to the sugar industry. Credit is the important instrument for cultivation of sugarcane, because 80% of the farmers depend on credit. If there is no assurance of credit market, the farmers could not cultivate sugarcane. Only 60% of the farmers want seeds from the company. According to some farmers, seeds are not important but If the company wants to increase the productivity of sugarcane cultivation then the company should provide the hybrids seeds. Here the extension service is not so significant. Only 5% of the farmers want extension service. According to farmers they know how to cultivate sugarcane better. Irrigation is very important for sugarcane cultivation. Here nearly 80% of the farmers want irrigation facilities. According to some farmers, they will shift to sugarcane cultivation if the irrigation facilities improve. That means if the company provides the pump sets and sanctions some credit for irrigation facilities, they can shift to sugarcane cultivation. Transportation is not also important for cultivating the sugarcane. Only 28% of farmer wants the transportation facilities. Most of the farmers also want the fertilizer from the company. It is because fertilizer occupies half of the cost of cultivation of sugarcane. Therefore, fertilizer is the major incentive for sugarcane cultivation.

Table 2
Productivity of Different Crops

Indicators		*Land (in acres)*	*Irrigated land (in acres)*	*Credit taken (in rupee)*	*Instruments*
Cultivators of sugarcane (31 HHs)		129.4 (54.16)	55.5 (76.60)	103000 (72.18)	58 (68.24)
	Average	4.17	1.79	3323	1.87
Non-cultivators of sugarcane (50 HHs)		109.5 (45.84)	16.95 (23.40)	39700 (27.82)	27 (31.76)
	Average	2.19	0.34	794	0.54
Total		238.9 (100)	72.45 (100)	142700 (100)	85 (100)
	Average	2.95	0.89	1761.7	1.04

Source: The Field Survey*Note:* Here the () shows the Percentages.

A contract will be successful when it works for a longer period. That means in short period of time it may not give a certain amount of profit to both parties but in the long run it may give certain profit to both parties. This is because in the short run; both the parties may not get the anticipated return from their investment. The present study found that 61 % of sugarcane cultivators are asking for long-term contract. This shows that the non-cultivators of sugarcane will shift to sugarcane cultivation and sugarcane cultivators will increase the area under sugarcane cultivation if the company runs for longer period. Among the sugarcane cultivators, some farmers are asking only for long-term contract. According to Arabinda shaoo a sugarcane cultivator, at the time of Ponni Sugar Company, he was cultivating eight acres of land for sugarcane but due to closure of the Ponni unit, he incurred great loss. He does not want therefore to expand the area under sugarcane cultivation and he will expand the area under sugarcane cultivation only after two/three years of running of the company. Most of the farmers want for a long-term contract because the sugarcane cultivation is a long run process. That means if a farmer starts to cultivate the sugarcane, it will go on for three to four years. There is no need of planting the cane every year. In addition, the contract should be for a long term for success.

Figure 1

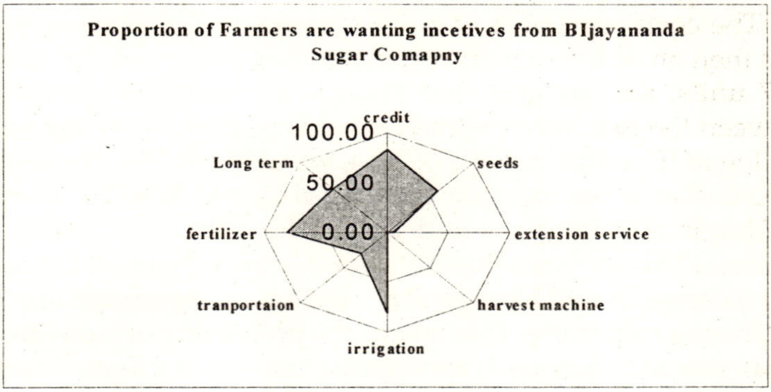

REGRESSION RESULT

From Equation seven the estimated result is

$$L_i = \left(\frac{P_i}{1-P_i}\right) = -3.217408 - 0.31924TL + 1.667564IL + 0.0005CR + 1.678948INST \quad (8)$$

(-4.115266) (-2.78700) (3.233484) (1.346976) (2.936597)

Each slope coefficient in this equation is a partial slope coefficient and measures the change in the estimated Logit for a unit change in the value of the given regressor (holding other regressors constant)

Here the constant term is -3.217408, which shows that if the constant term increases by a unit then on average the estimated Logit decreases by about -0.23 unit, suggests that there is a negative relationship between the two.

Thus, the TL coefficient of -0.31924, with other variables constant, means if TL increases by a unit on an average the estimated Logit decreases by about -0.32 units, suggests a negative relationship between the two. If we take the antilog of TL, coefficient of -0.313924 the study got 0.731036. This states that if we increase the total land by 1 unit, the Logit will decrease and hence the odds ratio will also decrease. This suggests that if the total land of a farmer increases, it will have a negative impact on the probability of cultivating sugarcane. This shows that the small farmers are cultivating sugarcane more proportionately than the large farmers. The small farmers are taking more risk than large farmers.

The coefficient of IL, 1.667564 means, if IL increases by a unit then on an average the estimated Logit increases by about 1.67 units, this suggest that there is a positive relationship between the two, other variables held constant. If we take the antilog of IL coefficient of 1.667564, we get 5.281499. The result reflects that if we increase the total irrigated land by 1 unit, the Logit will increase and hence the odd ratio will also increase. This suggests that if the total irrigated land of farmers will increase, it will have positive impact on the probability of cultivating sugarcane. That means the probability of cultivating sugarcane will increase if we increase the size of irrigated land.

However, the coefficient of CR is insignificant but coefficient of CR 0.0005 shows that if we increase by a unit, on average estimated Logit increases by about 0.0005 units, which shows that there is a positive relationship between these two. If we take the antilog of CR coefficient, 0.0005, we get 1.000499, which indicates that if CR increases by a unit the probability to cultivate sugarcane of the households will increase in relation to probability of not cultivating sugarcane.

The coefficient of INST 1.678 shows, if INST (plough, pump sets and cart) increases by a unit, an average estimated Logit increases by about 1.68 units, suggesting that there is a positive relationship between the two, where other variables hold constant. If we take the antilog of IL coefficient of 1.6789, it gets 5.342037. This shows that if we increase the total INST by 1 unit, the Logit will increase and the odd ratio will increase. This shows that if the INST will increases then the sugarcane production will increase.

CONCLUSION

Contract farming as it currently operates in India poses problems for many of the more disadvantaged groups in rural communities. Jayati Gosh, Devinder Sharma and others indicate that contract farming system has danger for small farmers. Despite these problems this is increasing. A joint venture will be successful when both the parties will get at least minimum profit by entering into this type of relation or game for a long period. Therefore, the primary precondition for any investment in contract farming must be profitable for a long period. The contract should specially be concerned with power relationships between big business and small growers and the need to strengthen the position of small growers and labourers (Gina Porter & Kevin Phillips-Howard, 1997). An industry or sponsor will invest in a particular project thinking it to be profitable; that means the sponsor will invest in such a market that has demand. In addition, market must have the capacity to remain profitable in the longer term, and the sponsor should analyze if production can be carried out profitably given the future expected price. The sponsor should know that farmers would obtain higher net incomes by entering into this system by cultivating certain crops rather than the other crops. Alternatively the company has to provide guaranteed, regular alternative incomes which will encourage farmers to make a long-term commitment. The sponsor needs to organize the supply of all necessary inputs for the farmers and for their own processing needs. All inputs should be identified and ordered well in advance either from local sources and/or from other sources.

In this context the present study found that the Company should handle the uncertainty existing in market in respect to sugarcane, the company should procure the whole amount of sugarcane produced by farmers at the right time. Company should assure price and inputs needed in the process of sugarcane cultivation to farmers. In return, farmers should handle the output uncertainty in sugarcane cultivation. And farmers should supply the sugarcane to the company. In an other sense for success of the agreement, the company should run for long period of time by which farmers can change the cropping pattern from paddy to sugarcane. The model shows that if the company will take some initiative in the area like credit, irrigation facilities and instruments, then the area under sugarcane cultivation may increase. From this process there can be a change in the cropping pattern of agriculture and farmer can earn more income, which leads to developing the economy.

NOTES

1. Risk is the fact the result of any action are not certain but may take more than one value, that the out come actual cause is not known, but it is expected that it will be determined as the result of a random drawing from set of possible outcomes where probability distribution is know. Uncertainty means an agent has lack of knowledge about the present fact and future possibility that the probability distribution are unknown.
2. Unlocking Growth in Andhra Pradesh
3. Contracting out Solution: Political Economy of Contract Farming in Indian Punjab By Sukhpal Singh 2002 World Development pp 1621-1638.
4. Bijayananda Sugar Cooperative Mill, different from the farmer's cooperatives, it is a member cooperative, where the group of member are promoters. It is an cooperative institution, where the members take decision about the company and all the members are not the farmers. Farmers' cooperative in an institution where, the farmers have active role for running the company and they contribute the shareholder for the company. In addition, the farmers are independently selecting their director of the company but in this case, farmers have no active participation for running of the company, only the farmers are supplying the sugarcane.
5. Ponni sugar (Orissa) Limited was an extension of Ponni sugar (erode) Limited in Tamil Nadu. Ponni Sugar Company was a private company that was successful in Tamil Nadu wants to replicate in the Orissa. Ponni sugar entered into Orissa in the year 1991 in response to an open invitation by the Government of Orissa to industrialists all over the country to establish sugar mills in Orissa. Due to some internal and external problem the company was closed in 2001. This company crushed the sugar cane only for five years. After the closure of the unit (Ponni Sugar and Chemical Orissa, Limited), 16th April 2001, Sri Jagneswar Babu, a local political leader and an ex-member of the Orissa Legislative Assembly took the lead to organize the farmers in

the form of a co-operative society to run this company and he changed the name as Bijayananda Sugar Co-operative Mill Ltd.

REFERENCES

Gulati Ashok, (2006), Fragmenting Bottom and Consolidating Top; India's Changing Food System and Implications for Small Holders, *Presented Paper in CESS Hyderabad,*

Banerjee Banerjee, *et al.* (2002), Strategy for Economic Reform in West Bengal, *"Economic and Political Weekly".*

Bharadwaj, K (1987), "Analysis of Agriculture-Industry Relation", *Economic and Political Weekly,* Vol. 22, No. 19 20 & 21.

Benziger, V. (1996), "Small Fields, Big Money: Two Successful Programs in Helping Small Farmers Make the Transition to High Value Added Crops". *World Development,* 24 (11), 1681-1693

Chand, Ramesh (1999a), "Emerging Crisis in Punjab Agriculture", *Economic and Political Weekly,* Vol. 34, No. 13, March 27-April, pp. A2-A10.

Clapp, R.A.J. (1988), "Representing Reciprocity, Reducing Domination: Ideology and the Labour Procurement in Latin American Contract Farming", *The Journal of peasant Studies, 16(1), 5-39*

Contract Farming Venture in India: A Few Successful Case, Vol. 1 March 2003, *Published by National Institute of Agricultural Extension Management (MANAGE).*

Eaton, Charles and Shepred, Andrew (2001), "Contract Farming: Partnership for Development", *Food and Agricultural Organization of United Nations*

Escorts gets into Contract Farming (2003), *April 10, Business Line.*

Glover, D. (1984), "Contract Farming and Smallholders Out grower Schemes in Less-Developed Countries", *World Development,* 12 (11/12), 1143-1157.

Glover, D. and Kluster, K. (1990), "Small Farmers, Big Business: Contract Farming and Rural Development", St. Martin's Press, New York.

Glover, D .J. (1987), "An Increasing the Benefits to Smallholders from Contract Farming: Problems for Farmers' Organizations and Policy Makers", *World Development,* 15(4), 441-448

Goldsmith, A (1985), "The Private Sector and Rural Development: Can Agribusiness Help the Small Farmer, *World Development,* 13 (10/11), 1125-1138

Key, N. and Runsten, D. (1999), "Contract Farming Smallholders, and Rural Development in Latin America: The Organization of Agro-Processing Firms and the Scale of Out Grower Production", *World Development,* 27(2), 381-401.

Linkages between Farmers and Enterprises for Contract Farming must be improved, *http://www.adb.org/Documents/news/vrm/vrm-200501.asp*

Pari Baumann (2000), Equity and Efficiency in Contract Farming Schemes: The Experience of Agricultural Tree Crops, *Working Paper 139 Overseas Development Institute 111 Westminster Bridge Road London SE17Jd UK.*

Porter, G. and Phillips-Howard, K. (1997), "Comparing Contacts: An Evolution of Contract Farming Schemes in Africa. *World Development,* 25(2), February 227-238.

Rangi, P.S. and Sidhu, M.S. (2000), "A Study on Contract Farming of Tomato in Punjab" *Agricultural Marketing,* 42(4), 15-23.

Report on Revival of Sugar Mill at Bolangir, *Prepared by IRMA (2003)*

S. Mahendra Dev and N. Chandrasekhra Rao (2005), Food Processing and Contract Farming in Andhra Pradesh; A Small Farmer Perspective, *Economic and Political Weekly,* June 25, 2005.

Singh, Sukhpal (2002), "Contracting out solutions: Political Economy of Contract Farming in Indian Punjab", *World Development,* Vol. 30(9), pp. 1621-1638.

Singh Sukhpal, (2005) Contract Farming system in Thailand *"Economic and Political Weekly".*

Singh Sukhpal (2005), Contract Farming for Agricultural Development: Review Theory and Practice with Special Reference to India, *CENTAD working Paper No. 2, An Oxfam GB Initiative, New Delhi.*

White, B. (1997), "Agro-industry and Contract Farming in Upland Java", *The Journal of Peasant Studies, 24(3), 100-136.*

"Orissa State Gazetteer, Vol. I, II&III" *Department of Revenue (Gazetteer Unit), Government of Orissa, Cuttack 1992, 200*

Agenda Notes for Districts Agricultural Strategy Committee Meeting of Bolangir District for Karif-2004: *Director of Agriculture Bolangir.*

Damodar N. Gujarati, Fourth Edition, Basic Econometrics, *Tata McGraw-Hill,*

15

MARKETED SURPLUS AND MARKET DEPENDENCE OF RICE, GREEN GRAM AND BLACK GRAM

SUDHAKAR PATRA & KABITA KUMARI SAHU

Food security has been the important thrust of agricultural research in recent times. The marketed and marketable surplus are important in the analysis of Food Security. The marketed surplus is the difference between the total agricultural outpout and home consumptions. There has been a lot of academic interest on marketed surplus of different agricultural crops. On the other hand the markateble surplus is the output net of seed, payments in kind and consumption at source. So the marketed surplus is marketable surplus plus distress sale. There are two concepts such as gross marketed surplus and net marketed surplus used in the study. The net marketed surplus is the gross marketed surplus minus the repurchase of food grains. The marketable surplus is of special importance because they are basic wage goods for the industrial sectors. The present field study is based on the marketed and marketable surplus of 150 farmer households in Haladia village of Cuttack District in Orissa. The variation in marketed and marketable surplus for different classes of households has significant influence on the availability of food grains. This paper presents the field research findings on marketed and marketable surplus with market dependence of rice, green gram & black gram. The study also investigates the important factors influencing the differences in marketed surplus.

MARKETED SURPLUS AND MARKETABLE SURPLUS

Though theoritically the distinction between marketed and marketable surplus is acknowledged, it has not always been maintained in several studies and the words are often used as interchangeable. The marketed surplus refers to the amount actually taken to the market whereas the marketable surplus is the output net of seed, payments in kind, and consumption at sources.

Marketed surplus means the actual quantity of the product marketed. It is found by substracting house consumption from total output of the product. But marketable surplus refers to the quantity that is available for sales. The cultivators' marketable surplus is estimated as the difference between the outputs and retention. Thus marketable surplus represents the residual product available with the farmers for disposal left after their genuine requirements of family consumption, payment of wages in kind, seed, rent, wastage and other contractual obligations such as loan repayments. So, marketed surplus is marketable surplus plus distress sales.

The difference between the marketed and marketable surplus arises not so much on account of at least two other factors. The poor farmers may sell out of distress and repurchase later, usually at higher prices. For such farmers, marketable surplus may very well be negative as they have to make net purchases to meet their consumption needs. The rich farmers,on the other, hand, may find it profitable to give loans in kind at exhorbitant rates of increase and on mortgage, and the quantities they len donot enter the market, For them, the marketable surplus could be larger than the marketed surplus.

Apart from these sectors, some differences between the marketed and marketable surplus can arise on purely statistical grounds where the sale or market disposals are not fully recorded but where the data base for estimating the marketable surplus is much more dependable.

Two concepts of marketed surplus can be used as gross and net. Gross marketed surplus refers to the actually marketed quantities and net marketed surplus is the gross marketed surplus minus repurchases of foodgrains. Net marketed surplus by itself does not become equal to marketable surplus.For analytical & policy purposes it is also important to distinguish between the gross marketable surplus and the net marketable surplus at the sectoral level, though it is not necessary at the house-hold level. From the point of the urban sector, it is the net marketable surplus, i.e. the marketable surplus net of the food purchases of all rural households that is not important. If procurement of marketable surplus is intended for distribution on both urban and rural sectors, the gross marketable surplus is relevant for indicating the procurable surplus. Analytically, the distinction between the gross and the net marketable surplus of the rural sector is important because even if under certain circumstances the gross quantity increases the net quantity need not.

HOUSEHOLDS AND LAND SIZE

There are 150 households selected on the basis of stratified random sampling for the present study in Haladia village which is an agriculturally developed village. The households are classified into six different classes according to land holdings. There are 42 households in class with 0 to 2 acres of land and 48 households in 2.1 to 4 acres of land. There are 28 households in 4.01 to 6.00 acres and 16 households in 6.01 to 10.00 acres of land. There are only 10 households with landholdings 10.01 to 15.00 acres and 6 households with above 15 acres of land. The adult per household are calculated which 5.54 for all classes is. The average household size is more in large farmers than small farmers. The household size varies between 5 to 7 when adult units are taken excluding the children. The owned area per adult is only 1.29 when all classes are taken together. It is only 0.23 in small farmer's class 0-2 acres and it increases upto 2.38 in large farmer class above 15 acres of land as shown below in Table 1.

Table 1
Household and Land Size under Study

Area in Acres	No. of House-hold	Adult Unit	Adult per House-hold	Owned Area per House-hold	Owned Area per Adult Unit
0-2.00	42	226	5.14	1.26	0.23
2.01-4.00	48	266	5.33	2.85	0.46
4.01-6.00	28	158	5.28	4.86	0.72
6.01-10.00	16	112	6.37	7.72	1.05
10.01-15.0	10	78	6.80	12.17	1.83
Above 15	6	50	6.83	20.98	2.38
All Classes	150	890	5.54	9.12	1.29

MARKETABLE SURPLUS OF RICE

Rice is the most important food grain produced in the study area of Orissa. The gross marketed surplus is 8.9k.g. for small farmers in class upto 2 acres of land. It is 17.8 kg for farmers coming under the class 2.01 to 4 acres It is 57.3kg. for the farmers in class 10 to 15 acres, and 61.8kg for the farmers in class above 15hec. when it is taken together it is 222.9kg for all class. The net marketed surplus is -1.9kg. for the classes of farmers between 0 to 2 acres it is -15.7kg. for farmers falling in the class 2.01 to 4 acres It is 48.8kg for farmers in class of 10 to 15acres and it is 73.6 kg for farmers in class above 15 acres. When all classes are taken together it comes to be 138.2kg

The percentage of grosss marketed surplus to the net agricultural output is 48.6% for the class of farmers of 0 to 2 acres 39.8% for the farmers in class from 2 to 4 acres It is as high as 53.7% for the size of class inbetween 10.1 to 15 acres When all classes are taken together it is 36.8%. The percentage of net marketed surplus to the total output is 27.3% for the farmers in class 0 to 2 acres It is 23.3% for the farmers in the class size of above 15 acres.

The marketable surplus is 10.8 for the farmers coming under the group 0 to 2 acres It is 187.4 for the farmers in class above 15 acres The marketable surplus as percentage of gross output for farmers coming under the size of class of 0 to 2 acres is 51.7%. It is 46.8% for the farmers in class 2.01 to 4 acres It is 71.5% for farmers in class 10 to 15acres and 75.9% for the

Table 2
Marketed Surplus of Rice by Land Holding Class (In KG)

Size Class in Acres Agricultural	Gross Marketed Surplus	Net Marketed Surplus	Percentage of GMS to net Output	Percentage of NMS to Net Agricultural Output	Marketable Surplus	MS ass Percentage of Gross Output	MS as percentage of NAO
1	2	3	4	5	6	7	8
0-2.00	8.9	-1.9	48.6	27.3	10.8	51.7	54.3
2.01-4.00	17.8	-15.7.	39.8	17.6	22.6	46.8	49.5
4.01-6.00	27.3	13.9	49.5	36.5	33.8	58.3	61.4
6.01-10.00	49.8	19.6	31.6	25.7	118.6	69.7	70.8
10.01-15.00	57.3	48.8	53.7	45.8	179.5	71.5	78.3
Above 15	61.8	73.6	24.6	23.3	187.4	75.9	83.6
All classes	222.9	138.2	36.8	29.6	552.7	69.3	73.5

Table 3
Marketed Surplus of Mung (Green Gram) by Land Holding Class (In KG)

Size Class in Acres Agricultural	*Gross Marketed Surplus*	*Net Marketed Surplus*	*Percentage of GMS to net Output*	*Percentage of NMS to Net Agricultural Output*	*Marketable Surplus*	*MS ass Percentage of Gross Output*	*MS as percentage of NAO*
1	*2*	*3*	*4*	*5*	*6*	*7*	*8*
0-2.00	1.2	-12.2	3.7	-29.6	-42.3	-73.6	-76.6
2.01-4.00	22.7	-15.3	18.6	-8.7	-16.5	-11.5	-13.7
4.01-6.00	21.2	-14.8	16.8	-6.3	-28.5	-20.6	-22.5
6.01-10.00	45.3	23.2	19.3	-3.4	53.2	12.3	16.8
10.01-15.00	78.5	61.8	27.5	22.6	131.1	32.5	37.3
Above 15	127.3	101.8	32.2	29.8	27.5	41.6	43.6
All classes	296.2	144.4	24.3	14.2	367.5	20.8	23.5

4 acres size class which is 23.3% in next class. It is only 5.9% in case of large farmers above 15 acres of land.

Table 5
Market Purchase of All Food Grains in a Year by Land Holding Class

Ownership Size Class in Acres	*Purchase Per Adult Unit (In K.g)*	*Purcentage of Total Consumption*	*Percentage of Gross Marketed Surplus*
1	*2*	*3*	*4*
Upto 2 acres	42	19.5	149.3
2.1 to 4.0	59	31.7	131.7
4.1 to 6.0	48	23.3	74.5
6.1 to 10.0	28	11.4	33.8
10.1 to 15.0	16	6.8	9.6
Above 15 acres	14	5.9	7.8
All size classes	31	14.1	29.1

The percentage of market purchase to gross marketed surplus shows wide variation among different classes. In case of small farmers the market purchase is more than the market surplus and percentage is 149.3% for farmers upto two acres of land . The percantage is 131.7% for farmers in class 2.1 to 4 acres It is as low as 9.6% for farmers within 10.1 to 15 acres and it is 7.8% for farmers above 15 acres of land when all classes of farmers are taken together the proportion of market purchase to marketed surplus is 29.1%. The analysis reveals that the farmers depend more on market purchase of food grains.

MARKET PURCHASE OF RICE

Rice is the most important food grain produced in the study area of Orissa. Many people eat rice three times a day and wheat consumption is very low in rural area. The market purchase of rice is 53kg per adult in small farmers class upto two acres which is 61 kg for next class. It is only 23 kg. for the class size 10 to 15 acres and 18kg for farmer in above 15 acres class. When all classes of farmers are pulled together per adult purchase of rice is 37kg in study area. The percentage of market purchase to total rice consumption is 25.3% when all size classes are considered together. It is highest 47.5% for small farmers upto 2 acres of land and the percentage decreases with increase in land size. It is only 6.1% in case of large farmer above 15 acres. The analysis reveals that the purchase of rice is less than 50% for all types of farmers and it decreases with increase in land size.

The study has also calculated percentage of rice purchase to total purchase of food grain for all types of farmers it is 13.1% which implies that the percentage of rice is relatively less compared to other food grains. The farmers upto 2 acres of land have 11.2% of rice purchase to total purchase of food grain. It is highest 24.3% for farmers with 10 to 15 acres of land and it is lowest 9.7% for farmers above 10 acres.

Table 6
Market Purchase of Rice in a Year by Land Holdind Class

Ownership Size Class in Marketed Acres Rice	*Purchase Per Adult Unit (in K.g)*	*Purcentage of Total Rice Consumption*	*Percentage of Total Purchase of Food Grain*	*Percentage of Gross Marketed Surplus*
1	*2*	*3*	*4*	*5*
Upto 2 acres	53	47.5	11.2	41.3
2.1 to 4.0	61	43.2	10.3	55.8
4.1 to 6.0	57	39.8	11.1	28.3
6.1 to 10.0	38	26.2	13.6	9.4
10.1 to 15.0	23	22.6	24.3	10.5
Above 15 acres	18	6.1	9.7	5.8
All size classes	37	25.3	13.1	17.2

The percentage of market purchase of rice to gross marketed surplus of rice is 41.3% for farmers within 2 acres of land. It is only 5.8% for large farmers above 15 acres of land . The percentage of market purchase to marketed surplus of rice is highest 55.8% for farmers 2.1 to 4 acres and there after it decreases for farmers of higher land holding class. It is found in study area that the farmers cultivate rice in maximum area of their gross cropped land and they depend on market purchase for other types of food grains.

Table 7
Market Purchase of Blackgram in a Year by Land Holding Class

Ownership of Size Class in Acres	*Purchase Per Adult Unit (Per K.g)*	*% of Total Black Gram Consumption*	*% of Total Purchase of Food Grain*	*% of Gross Marketed Surplus of Black Gram*
1	*2*	*3*	*4*	*5*
Upto 2 acres	24.2	17.3	47.6	657.2
2.1 to 4.0	31.7	26.8	44.8	149.6
4.1 to 6.0	29.3	20.6	59.3	151.8
6.1 to 10.0	12.8	7.3	42.4	36.3
10.1 to 15.0 acres	6.5	3.4	41.8	9.2
Above 15 acres	2.6	2.1	22.6	2.9
All size classes	15.3	9.6	47.4	34.6

The purchase of blackgram per adult unit is 24.2kg for farmers in the class 0 to 2 acres It is 31.7kg for farmers coming under the class of 2.01 to 4 acres It is 6.5kg per adult unit in the household coming in the class of 10.01 to 15acres it is 2.6kg for farmers coming in the class of above 15 acres

The percentage of total blackgram consumption is 17.3% for farmers coming in the class 0 to 2 acres, it is about 26.8% for the farmers coming in the class 2.01 to 4 acres, whereas the percentage of total black gram consumption is 7.3% for the farmers in the class 6.01 to 10 acres It is as low as 3.4% and 2.1% for the farmers coming in the class 10.01 to 15 acres and above 15 acres respectively. But when all classes taken together it comes to be 9.6%.

The percentage of total food grain purchase is 47.6% for the class of farmers coming in the group 0 to 2 acres It is 44.8% for the farmers coming in the class 2.01 to 4 acres whereas for the farmers in the group of 10.01 to 15 acres it is 41.8% and it is 22.6% for those who came under the class of above 15acres when all classes taken together it comes to be 47.4%.

The percentage of gross marketed surplus of black gram is 675.2% for the farmers in the class 0 to 2 acres It is 149.6% for the farmers in the class 2.01 to 4acres where as it is 151.8% for the farmers coming in the group between 4.01 to 6 acres It is 36.3% and 9.2% for the farmers who come under the group in between 6.01 to 10acres and 10.01 to 15 acres resectively It is very low (2.9%) for the farmers in the group of above 15acres when all classes are taken together it is 34.6%.

So, the above table reveals that there is a considerable variation in the purchase of black gram for the farmers those who come under the class of 0 to 2 acres and for those farmers who come under the class above 15. Same is also the result in case of percentage of total blackgram consumption and percentage of gross marketed surplus of black gram.

The purchase of green gram per adult unit is 7.2kg and 13.3kg for the farmerss coming in the class from 0 to 2 acres and 2.01 to 4 acres Itis 6.9 kg for the farmers in the group 4.01b to 6 acres and 4.8kg. for the farmers in the class above 15acres. When all classes taken together, it comes to be 5.7kg. The percentage of total green gram consumption is 13.4% and 33.2% for the farmers coming in the class of 0 to 2 acres and 2.01 to 4 acres It is 16.8% for the farmers in the class between 4.01 to 6acres It is 10.3% for the farmers in the class of 6.01 to 10 acres and the lowest 1.4% for the group of farmers coming under the class 11.7% to 15 acres It is the highest 12.8% for the farmers in the group of above 15 acres When, all classes are taken together it comes to be 19.6%. The percentage of total food grain purchase is 18.4% and 21.7% for the farmers in the group 0 to 2 acres and 2.01 to 4 acres It is the 10.01 to 15 acres and the highest 36.4% for the farmers who are falling in the class of above 15 acres When all classes are taken together, it comes to be 19.6% The percentage of grosss marketed surplus of green gram is 152.8% and the highest 287.4% for the farmers in the classs 2.01 to 4 acres The percentage of gross marketed surplus of green gram is 38.1% and 42.7% for those farmers who fall in the class category of 4.01 to 6 and 6.01 to 10 acres respectively. It is the lowest, 2.3% for the farmers in the class 10.01 to 15acres and it is 13.8% for the farmers in the class category of above 15acres But when all classes taken together it comes to be 29.9%.

PAYMENT AND RECEIPTS IN KIND

In rural area the barter system still prevails in limited scale and non-monetary transactions take place. The food grains are also given in the form of loans and gifts by the friends and

marketed surplus of price is 2229kg only. The percentage of gross marketed surplus of rice to the net agicultural output is 48.6% for the marginal farmers. The marketable surplus in green gram production is negative for farmers with less than 2acres of lands. The percentage of gross marketed surplus to net agricultural output is higher than the percentage of net marketed surplus to net agricultural output. In case of black gram the percentage of gross marketed surplus to net agricultural output is 16.2% for marginal farmers below 2 acres of land the behaviour of marketable surplus and the relationship with output is discontinuous and non-linear .

The market dependence refers to purchase of food grains by the people from the market. The per adult unit market purchase is 31kg for all size classes in the study village. The percentage of market dependence to gross marketed surplus shows wide variation among different classes. The rice consumption is very high in rural Orissa. Compared to wheat. The market purchase of rice per adult unit is 37kg. Which is 25.3% of total rice consumption. The market purchase of black gram is 15.3kg. which is 9.6% of total black gram consumption in the study village. The market purchase of green gram is 5.7kg per adult unit which is 12.8% of total gren gram consumption. The payment in kind of green gram is highest for the small and marginal farmers and lowest for those farmers who hold land area above 15 acres. It is observed from the analysis that payment in kind of black gram is more than the green gram for the farmers. The market purchase per adult unit is highest for farmers in the category of 2.1 to 4 acres of land whereas it is lowest for large farmers above 15 acres of land.

REFERENCES

Babatunde R. O. and Others (2008), Determinants of Vulnerability to Food Insecurity, *Indian Journal of Agricultural Economics,* Vol. 63, No. 1, Jan-March, pp. 116-125.

Khokhar K. S. (2006), Meeting Agricultural Challenges, *Kurukshetra,* Vol. 54, No. 5, pp. 43-44.

Kumar A. and Mrutyunjay (1989), Price Policy and Marketed Surplus of Paddy and Wheat in India, *Indian Journal of Agricultural Economics,* Vol. 44, No. 4, Oct–December.

Nanavaty R. (2008), Towards Livelihood Security, *Yojana*, Vol. 52, Jan, pp. 21-24.

Radhakrishna R. (2002), Agricultural Growth, Employment and Poverty–A Policy Perspective, *Economic & Political Weekly*, Vol. 30, 19th Jan, pp. 243-245.

Tripathy K. K. & Jain S. K. (2007), Agriculture in the Post Reform Era, *Kurukhetra*, Vol. 55, No. 9, P. 3-10.

INDEX